# The God-List in the Treaty between Hannibal and Philip V of Macedonia

THE JOHNS HOPKINS NEAR EASTERN STUDIES
Hans Goedicke, General Editor

Hans Goedicke, *The Report about the Dispute of a Man with His* Ba
J. J. M. Roberts, *The Earliest Semitic Pantheon*
John D. Schmidt, *Ramesses II: A Chronological Structure for His Reign*
David Lorton. *The Juridical Terminology of International Relations in Egyptian Texts through Dyn. XVIII*
Bernard Frank Batto, *Studies on Women at Mari*
Hans Goedicke, *The Report of Wenamun*
Hans Goedicke and J. J. M. Roberts (eds.), *Unity and Diversity: Essays in the History, Literature, and Religion of the Ancient Near East*
Patrick D. Miller, Jr., and J. J. M. Roberts, *The Hand of the Lord: A Reassessment of the "Ark Narrative" of 1 Samuel*
Thomas W. Mann, *Divine Presence and Guidance in Israelite Traditions: The Typology of Exaltation*
Hans Goedicke, *The Protocol of Neferyt*
Jack M. Sasson, *Ruth: A New Translation with a Philological Commentary and a Formalist-Folklorist Interpretation*
Michael L. Barré, *The God-List in the Treaty Between Hannibal and Philip V of Macedonia: A Study in Light of the Ancient Near Eastern Treaty Tradition*
Jerrold S. Cooper, *The Curse of Agade*
James Miller Lindenberger, *The Aramaic Proverbs of Ahiqar*

# THE GOD-LIST IN THE TREATY BETWEEN HANNIBAL AND PHILIP V OF MACEDONIA:

## *A Study in Light of the Ancient Near Eastern Treaty Tradition*

Michael L. Barré



The Johns Hopkins University Press
*Baltimore and London*

Printed in the United States of America

The Johns Hopkins University Press, Baltimore, Maryland 21218
The Johns Hopkins Press Ltd., London

**Library of Congress Cataloging in Publication Data**

Barré, Michael L.
The God-list in the treaty
between Hannibal and
Philip V of Macedonia.

(Near Eastern Studies)
Revision of thesis (Ph.D.)—Johns Hopkins
University, 1978.
Bibliography: p. 189
Includes index.
1. Gods. 2. Near East—Foreign relations—
Treaties. I. Title. II. Series.
BL473.B35 1983 291.2'11 82–13961
ISBN 0–8018–2787–6

*To my parents,*
*Lee John Barré*
*and*
*Mary Lonergan Barré,*
*in loving gratitude*

CONTENTS

## PREFACE

The present work represents an extensive revision of my doctoral dissertation, submitted to The Johns Hopkins University in 1978: *The God-List in the Treaty Between Hannibal and Philip V of Macedonia: A Study From the Viewpoint of the Ancient Near Eastern Treaty Tradition.* In the first chapter, I have added a section explaining the purposes, methodology, and some of the terms used throughout the work. The most substantial revision may be found in Chapter III. The material in this chapter has been rearranged to some extent and expanded to include a discussion of a greater number of Neo-Assyrian treaty god-lists. In Chapter IV the introductory section has been rewritten as has part of the discussion of "Zeus." I have expanded Appendix I to include six new charts and have appended a number of footnotes. A second appendix has also been added--actually an excursus--treating the question of foreign gods in treaty god-lists from Mesopotamia. Finally, as a result of the additional material in the body of the work itself, necessitating the addition of a number of footnotes, the List of Works Cited has been expanded.

I am grateful to a number of individuals whose teaching, assistance, or inspiration have played an important part in the writing of this dissertation. I express sincerest thanks first of all to Delbert R. Hillers, who suggested the topic of this dissertation and who was kind enough to serve as adviser and first reader. His characteristically careful and meticulous scholarship has been a model for me in the course of my own work. To J.J.M. Roberts, who served as second reader, I am deeply indebted, especially for his willingness to share so generously his time and learning. I owe a special word of thanks to Minor M. Markel for his gracious assistance in matters pertaining to classical studies. I wish also to express my gratitude to P. Kyle McCarter, Jr., for the investment of his time and interest during the process of revision; and to my friend and colleague, John S. Kselman, for his many valuable comments and suggestions.

At this time I would like to acknowledge a special debt of gratitude to Hans Goedicke, General Editor of The Johns Hopkins Near Eastern Studies Series, for his solicitude and encouragement and for his role in accepting the present work for publication in that distinguished series.

Finally, I am obligated to my superiors in the Society of St. Sulpice for providing me the time as well as the personal and financial support to pursue my graduate studies; without their assistance the present work could not have been written.

## LIST OF ABBREVIATIONS

The abbreviations used in this work follow those given in the "Instructions for Contributors" in the *Journal of Biblical Literature* 95 (1976) 335-46, with the following additions or changes:

| | |
|---|---|
| *AAA* | *Annals of Archaeology and Anthropology* (Liverpool). |
| *ABL* | *Assyrian and Babylonian Letters Belonging to the K. Collection of the British Museum* (ed. R. F. Harper). |
| *ABoT* | *Ankara Arkeoloji Müzesinde bulunan Boğazköy Tabletleri* (Istanbul). |
| AIPHOS | Annuaire de l'Institut de Philologie et d'Histoire Orientales et Slaves (Brussels). |
| *AJP* | *American Journal of Philology* (Baltimore). |
| A/M | The treaty between Assur-nirari V of Assyria and Mati'ilu of Arpad (ca. 750 B.C.). |
| A/Q | The treaty between Assurbanipal of Assyria and the Qedar tribe (mid seventh century B.C.). |
| *AT* | *The Alalakh Tablets* (ed. D. J. Wiseman). |
| *BAR* | *Biblical Archaeology Review* (Washington, D.C.). |
| *BBS* | *Babylonian Boundary-Stones and Memorial-Tablets in the British Museum* (ed. L. W. King). |
| *BRM* | *Babylonian Records in the Library of J. Pierpont Morgan* (ed. A. T. Clay). |
| CH | Codex Hammurapi. |

| | |
|---|---|
| *CHD* | *The Hittite Dictionary of the Oriental Institute of the University of Chicago* (ed. H. G. Güterbock and H. A. Hoffner). |
| DN | Divine name. |
| E/B | The treaty between Esarhaddon of Assyria and Ba$^{c}$al of Tyre (677 B.C.). |
| EPROER | Etudes préliminaires aux religions orientales dans l'empire romain (Leiden). |
| F | *Staatsverträge des Hatti-Reiches in hethitischer Sprache* (ed. J. Friedrich):<br>F1: Mursilis II and Duppi-Tešub<br>F2: Mursilis II and Targasnallis<br>F3: Mursilis II and Kupanta-KAL<br>F4: Mursilis II and Manapa-U<br>F5: Muwatallis and Alaksandus<br>F6: Suppiluliuma I and Ḫuqqanas. |
| GN | Geographical name. |
| Gsell 4 | S. Gsell, *Histoire ancienne de l'Afrique du Nord:* Vol. 4: *La Civilisation carthaginoise.* |
| H/P | The treaty between Hannibal of Carthage and Philip V of Macedonia (215 B.C.). |
| KBo | *Keilschrifttexte aus Boghazköi* (Leipzig/Berlin). |
| KUB | *Keilschrifturkunden aus Boghazköi* (Berlin). |
| LAPO | Litteratures anciennes du Proche-Orient (Paris). |
| *MIO* | *Mitteilungen des Instituts für Orientforschung* (Berlin). |
| MRS | Mission Ras Shamra (Paris). |
| M/Š | The treaty between Marduk-zakir-šumi I of Babylon and Šamši-Adad V of Assyria (ca. 820 B.C.). |
| Obv | Obverse: corresponds to *Vs. (Vorderseite)* in German and *R$^{o}$* (recto) in French editions of cuneiform texts. |
| PBS | Publications of the Babylonian Section, University Museum, University of Pennsylvania. |
| *PDK* | *Politische Dokumente aus Kleinasien* (ed. E. F. Weidner). |

| | |
|---|---|
| *Pr. Ev.* | Eusebius, *Praeparatio evangelica.* |
| Rev | Reverse: corresponds to *Rs. (Rückseite)* in German and *V*° (verso) in French editions of cuneiform texts. |
| *RHA* | *Revue hittite et asianique* (Paris). |
| RS | Ras Shamra text, inventory number. |
| SANE | Sources and Monographs on the Ancient Near East (Malibu, Calif.). |
| Senn | The Sennacherib treaty-fragment (VAT 11449). |
| Sf1 | The treaty between Bir-Ga'yah of KTK and Mati$^{c}$el ( = Mati'ilu) of Arpad (ca. 750 B.C.). |
| S/N | The treaty between Sin-šum-lišir (?) of Assyria and Nabu-apli-iddina et al. (ca. 625 B.C.). |
| StBoT | Studien zu den Boğazköy-Texten (Wiesbaden). |
| StOrFen | Studia Orientalia edidit Societas Fennica (Helsinki). |
| StSem | Studi semitici (Rome). |
| *TDOT* | *Theological Dictionary of the Old Testament* (ed. G. J. Botterweck and H. Ringgren). |
| *THAT* | *Theologisches Handwörterbuch zum Alten Testament* (ed. E. Jenni and C. Westermann). |
| VAT | Inventory numbers of tablets in the Staatliche Museen in Berlin. |
| VTE | The vassal-treaties between Esarhaddon of Assyria and (chiefly) his Median vassals (672 B.C.). |
| W | *Politische Dokumente aus Kleinasien* (ed. E. F. Weidner):<br>W1: Suppiluliuma I and KURtiwaza<br>W2: KURtiwaza and Suppiluliuma I<br>W3: Suppiluliuma I and Tette<br>W4: Suppiluliuma I and Aziru |
| *WdM* | *Wörterbuch der Mythologie:* Vol. 1: *Götter und Mythen im Vorderen Orient* (ed. H. W. Haussig). |

Chapter I

# INTRODUCTION

## A. General Introduction

In the year 216 B.C., the third year of the Second Punic War, Hannibal inflicted a crushing defeat on the Roman forces at Cannae on the Italian peninsula. It is generally conceded that this victory convinced Philip V of Macedonia to cast his lot with Carthage in the war.[1] Consequently Philip sent an emissary to Hannibal, a certain Xenophanes, to conclude an alliance with the Carthaginian general. This alliance was formalized in an oath, or more precisely, a treaty[2] between the two parties during the summer of 215 B.C. But upon setting sail to return to Philip, Xenophanes and several Punic representatives were intercepted by the Roman fleet; they (and the documents) were then conveyed to the Senate in Rome.[3]

According to E. J. Bickerman, the Greek historian Polybius (ca. 200-120 B.C.) "was allowed to make copies of these documents in the Roman archives"[4] several generations after the writing of the treaty. Polybius enjoyed access to official records in both Greece and Rome and frequently made use of them.[5] That the present document was not composed by Polybius is evident from the fact that its style and vocabulary differ significantly from his own.[6] His *Histories* have not survived in their entirety, but fortunately the text of the treaty (7.9) has been preserved in a tenth- or possibly early eleventh-century Byzantine manuscript now in the Vatican *(Vaticanus Urbinas Graecus 102).*[7] Thus there has come down to us a text which is a "verbatim"[8] copy of the third-century B.C. document in the Roman archives.

Bickerman has demonstrated a further interesting aspect of the document copied by Polybius: it is a literal translation of the official Punic version of the treaty.[9] This is apparent from a number of Semitic expressions and syntactical constructions in the text.[10]

This, then, is in some respects a unique document: not only does it represent the only extant text of a Carthaginian treaty, but it is the closest thing we have to a treaty in Punic, the language of Carthage. The document is important for the study of treaties from the Semitic world or, in a broader perspective, from the ancient Near East. It is the latest such treaty to have survived--hence its importance for our understanding of the history of the treaty tradition in this area of the ancient world.

There is one section of this treaty, however, that appears to be particularly fruitful for the study of the religious and legal traditions of the ancient Near East--the list of divine witnesses. This brief section of the document has received more attention than any other. One reason for this apparently disproportionate interest lies in the fact that here we seem to have a listing of the gods of the "official pantheon"[11] of Carthage. Aside from this text our first-hand knowledge of this pantheon is quite fragmentary. There are numerous votive inscriptions from Carthage and the rest of the Punic world, it is true; but these rarely list more than several deities. Because of this fact, and since such inscriptions belong to the category of cult or private devotion rather than affairs of state, they tell us little or nothing about the official ranking of the high gods. At present, excavation has been resumed at Carthage, so that the near future may bring welcome additions to our knowledge in this area.

Another reason for the importance of the list is that it holds the promise of containing a far greater proportion of features parallel to those of the ancient Near Eastern treaties than the remainder of the document. There are several reasons for this assumption. First of all, this section is the most religious in tone and content of the whole text; and it is well known that religious terminology and formulas often exhibit considerable conservatism. Secondly, to one familiar with the treaty formularies of the ancient East this part of the text appears even on superficial examination to contain terms and phrases that correspond strikingly to those in more ancient treaties.

Some parallels between the Hannibal/Philip V treaty (H/P) and earlier ones from the ancient East have already been noted in Bickerman's studies. He concludes that "the structure and terminology of Hannibal's Oath roughly correspond to the language and composition of the Near Eastern treaties of the second millennium B.C."[12] But the extent of these parallels has never been adequately explored. In order to do this one must be prepared to deal extensively with all of the relevant treaty material, Akkadian, Aramaic, and Hittite, as well as related documents in other languages.

The scope and concern of this work, then, will be limited to a discussion of the god-list in H/P and relevant parallels from the Near East. That is to say, the present work will in no way concern itself with the issues of the military or political career of Hannibal or indeed with any historical personality as such. Nor will it deal directly with any aspect of the Punic wars.

In Chapter II we shall briefly discuss previous attempts to understand the structure and content of this god-list. Chapter III will be devoted to a treatment of the position of god-lists within ancient treaties as well as their structure and content; a number of these lists are reproduced in Appendix I. A detailed discussion of them will be necessary, since to date no thoroughgoing discussion of all the treaty god-lists exists. In the course of comparing their structure with that of H/P we shall have occasion to analyze its structure as well. Finally, Chapter IV will consist of a line by line commentary on the god-list. In the case of the first nine divine names there is an added problem: the names are those of Greek deities, but they are generally thought to conceal Punic gods. The difficult question is, which Punic deities lie behind the Greek divine names? Obviously this question is very important for our understanding of the "official pantheon" of Carthage at the time of Hannibal. It will therefore be necessary to devote considerable space to it. Since the problem of identifying these gods is a notoriously difficult one, we cannot rely on the treaty tradition alone but must make use of whatever sources and evidence are available. Next, the remaining lines of the god-list will be discussed, again with reference to the treaty tradition of the Near East.

In his first article on this treaty Bickerman noted, "This exceptional document has not received the attention it merits."[13] It must be admitted that scholars in the field of Near Eastern studies have been particularly at fault here. Although classicists and classical historians (to whom we must credit virtually all of the groundwork in the interpretation of our text) have pointed to the need for orientalists to bring their expertise to bear on matters relating to this and other Near Eastern treaties,[14] the response of their colleagues in the oriental field has been weak indeed. The present paper, then, may be viewed as an effort to give some merited attention to at least one section of this long-neglected document. It is also hoped that our study may provide a stimulus for further research into this unique treaty.

## B. Purpose, Methodology, Explanation of Terms

The purpose of the present work is threefold. The primary aim is to elucidate the god-list in H/P. This will involve an analysis of form as well as content. How is the god-list arranged? Who are the gods named in the list? Why are these particular deities invoked? What is the precise meaning of the phrases such as *theōn tōn systrateuomenōn*, *theōn pantōn hosoi katechousi GN*, and *hosoi tines ephestēkasin epi toude tou horkou?* Since there are no other extant Punic treaties with which to compare H/P, one can elucidate this treaty god-list only by comparing its arrangement and content with those of similar documents from other ancient civilizations. In the course of such a comparison one sheds light not only upon H/P itself but at the same time upon the relationship between this god-list and more ancient ones. Hence a secondary purpose is to examine the formal and material parallels to other god-lists from antiquity. The emphasis here will be on parallels from the ancient Near Eastern treaty tradition. The reason for this emphasis is first of all that research into our treaty in the light of similar Greek documents has already been done by classical scholars such as Bickerman, F. W. Walbank, L. F. Benedetto, and others. In particular the provocative conclusions of Bickerman regarding the connection of H/P with ancient Near Eastern treaty traditions[15] have set the stage for an investigation of this document in the light of these traditions. Moreover, the present writer is not a classicist and hence must leave an exhaustive investigation of parallels with the Greek treaty tradition to those at home in that field. However, relevant data from ancient Greek treaties will be included in the discussion where pertinent. Finally, a tertiary aim is to analyze the form and content of treaty god-lists from the ancient Near East. It is obvious that one can draw valid and meaningful parallels in form and content from the god-lists of Near Eastern treaties only to the extent that one understands the structure and content of those documents. Buch an analysis becomes a necessary undertaking for the present work when it is realized that very little has been done in this area to date.

As is apparent from the foregoing statement, a basic premise of our work is that H/P bears closest affinities to the treaties of the ancient Near East. This is not a gratuitous assumption. In the first place, such a connection has already been noted by Bickerman.[16] Secondly, a number of enigmatic phrases in the document cannot be explained on the basis of Greek treaties but find almost verbatim equivalents in Near Eastern treaty formularies. Since we are merely setting forth a premise here, it would be out of place to provide extensive documentation at this time. One striking parallel will be given here.

H/P begins as follows:

> Ὅρκος, ὃν ἔθετο Ἀννίβας ὁ στρατηγός . . .
> πρὸς Ξενοφάνη Κλεομάχου Ἀθηναῖον πρεσβευτήν
> . . . Ἐναντίον Διὸς καὶ Ἥρας καὶ Ἀπόλ-
> λωνος . . . (Polybius, *Histories* 7.9.1-2).
>
> The "oath" which Hannibal the general . . .
> made [lit., "set"] . . . with the ambassador
> Xenophanes the Athenian, son of Cleomachus
> [representing Philip V] . . . in the presence
> of Zeus and Hera and Apollo. . . .

No known Greek treaty begins in this way.[17] Moreover, Bickerman describes these lines as "strange and ungrammatically assembled"[18] and characterizes the use of *etheto* here as "a solecism."[19] Although Walbank points to several examples of *tithēmi* used in association with *horkos*,[20] the usage remains "unusual" even in his view.[21]

Compare, however, the first lines of a seventh-century B.C. Neo-Assyrian treaty:

> *a-de-e ša* $^{m}$*Aš-šur*.PAP.AŠ XX ŠÚ XX KUR *Aš-šur*
> . . . TA* $^{m}$*Ra-ma-ta-a-a* EN URU $^{uru}$*Ú-ra-ka-za-ba-nu*
> . . . *iš-ku-nu-ni ina* IGI $^{d}$SAG.ME.GAR $^{d}$*Dil-bad*. . . .[22]

> The treaty [lit., "treaty-oaths"] which Esarhaddon, king of the world, king of Assyria . . . made [lit., "set"] with Ramatay(a), city-ruler of Urakazabanu, in the presence of Jupiter, Venus. . . .

Other ancient Near Eastern treaties begin similarly.[23]

What type of ancient Near Eastern material may be used as a comparative basis for understanding the structure and content of the H/P god-list? The answer must be based on the nature of treaties themselves and, more particularly, of treaty god-lists. First of all, a treaty between nations is by definition not a private or casual undertaking but rather an official act of state. That H/P was understood by its drafters to be an official document and not a private pact between Hannibal and a foreign party--regardless of what the Carthaginian senate's official reaction to the treaty may have been --is clear from the fact that the treaty was concluded not only in the name of Hannibal but also in the name of "Mago, Myrcan, Barmocar, and all other [?] Carthaginian senators present with him, and all Carthaginians serving under him. . . ."[24] Hence valid comparison with texts from other areas or periods of the ancient Near East demands that data used for direct comparison be drawn from official documents. Since virtually all Near Eastern powers were monarchies in ancient times, "official documents" are equivalent to "royal documents" of a public nature. Secondly, a treaty concluded between powers has some of the characteristics of a legal document. Such a document would specify certain obligations and agreements which would be considered binding upon the parties involved. More particularly, the concept of a list of divine witnesses or guarantors (such as is found in all treaties from the Near East) is obviously modelled on that of human witnesses to legal transactions. Thus some aspects of ancient Near Eastern legal tradition may be brought to bear on the understanding of the H/P god-list. For example, *kudurrus* or "boundary stones" from ancient Babylon are at once legal and royal inscriptions;[25] moreover their affinities to treaties are widely recognized.[26]

As regards the phenomenon of the god-list itself, an important distinction must be made. A "god-list" may be defined as any consecutive series of three or more divine names appearing in a written document. But which deities are included in a particular list and how they are arranged depends to a large extent on the type of document in which they are found. W. G. Lambert has recently drawn attention to "the separation of state and private religion" in Mesopotamia.[27] There a separation obtained not only between state and private religion but in some instances between national and local cults.[28] The importance of this observation for the study of god-lists lies in the fact that a deity who may be important in the "state" or "official" religion may be considered minor in the "popular" religion and vice-versa. By extension, god-lists drawn from documents associated with the state religion are likely to differ in form and/or content from similar lists in writings reflecting popular religion. Since this point will prove methodologically important in the course of the present study, several examples will be given here.

The principle outlined above would appear to be true throughout the history of the ancient Near East. In Mesopotamia, for instance, "it is often pointed out that certain great deities, venerated throughout the entire land, as for example Utu or Nanna, held no place of importance in the cult of Lagaš."[29] Turning to Hatti, we note that in an official document (the Hittite "military oath") the moon-god plays "a central role as a retributive deity, in surprising contrast to his minor importance in the Hittite pantheon outside the oath."[30] Conversely, "Many deities who had a significant place in Hittite religion . . . are unaccountably omitted from the treaties. . . ."[31] Above all, one must resist the temptation, succumbed to by all too many students of ancient Near Eastern religion, to equate popularity in the cult with rank in the official or state pantheon. This caveat applies to West Semitic deities as well, as J. G. Février has noted: "Popularity and official rank are two different things in religious as well as political life."[32] In summary, one is ill-advised to draw conclusions about rank in the official (state) pantheon from the attested popularity of a deity in cultic texts, votive inscriptions, or onomastica.[33]

Thus in the present study texts from Near Eastern sources adduced as direct parallels to the H/P god-list will be restricted to legal and/or official (i.e., royal) inscriptions. An exception to this restriction will be the material adduced as relevant for the identification of the divine names in the list. It will be shown in the course of the discussion that the Greek divine names answer to Punic gods. The question of which

Punic deities are involved can be answered only in part from structural considerations; to a larger extent their identification must be based on a variety of literary and inscriptional material from the Mediterranean area.

We shall now define some of the more important terms used throughout this study. The term "pantheon" originally meant a temple dedicated to "all the gods." It is commonly used now, however, to mean "the gods of a people; esp.: the officially recognized gods."[34] Thus "pantheon" (without further qualification) will refer to the sum total of deities of a particular people, although this does not imply that all of them were venerated in the cult at all times. We shall therefore speak of the Carthaginian pantheon, the Tyrian pantheon, the Babylonian pantheon, etc. "Official pantheon" will mean the totality of gods listed in the official inscriptions of a particular nation. Strictly speaking, no single inscription contains the official pantheon but only a more or less comprehensive representation thereof. Obviously, too, the concept is somewhat fluid. For example, in Assyrian inscriptions one finds basically the same divine names in god-lists over the centuries; yet even some of the highest deities (although never the supreme god, Assur) may occasionally be omitted.

Since gods in the world of the ancient Near East were modelled on human (especially royal) prototypes, there was a certain hierarchy among them as in human society. We shall use the term "high gods" to denote those deities who were considered important enough to be included in the official pantheon. These gods may be determined in various ways. (1) In the Sumero-Akkadian pantheon each of the highest deities was assigned a particular number, which sometimes was used as a logogram to write the god's name. The numbers began with 60 (the Babylonian numerical system was sexagesimal), denoting the supreme god, as follows:[35]

| | |
|---|---|
| Anu | 60 |
| Enlil | 50 |
| Ea | 40 |
| Sin | 30 |
| Šamaš | 20 |
| Adad | 10. |

These six gods, the first three and the last three respectively, formed the two great "triads" of the Sumero-Akkadian pantheon.[36] The great goddess Ištar was also ranked among the highest deities; her number was

15, half that of her father, Sin.[37] In the Babylonian and Assyrian official pantheons, Marduk and Assur respectively were ranked above Anu. In Mesopotamia other deities (e.g., Nabu, Nergal, Ninurta, Nusku, Gula, etc.) were frequently listed just after these highest gods in official lists and hence are also included among the "high gods." Thus in Mesopotamia "high gods" (as used throughout this work) are equivalent for all practical purposes to the "gods of the official pantheon"; these, in turn, are equivalent to "oath-gods," that is, the deities invoked in treaties and other official ceremonies involving an oath. In Hatti, on the other hand, a number of gods were invoked who are clearly not "high gods"--for example, the "olden gods."[38] (2) Rank among the gods is regularly reflected in official god-lists. For example, the sequence Anu through Adad as shown above is found in the treaties as well as in royal inscriptions.[39] It is important to note this fact, namely that the ranking of the high gods (and especially the first position of the supreme state god) is found not only in treaties but in other types of official inscriptions as well, and that this practice obtained not only in Mesopotamia but also in other areas of the ancient Near East. In Hatti the several gods listed in the annals of Mursilis (II) as his helpers in battle[40] are basically the same (with two exceptions)[41] as the deities invoked in Hittite treaties; the hierarchy of gods listed is also the same in both types of official documents. This consistency of god-lists in royal inscriptions is found also in extreme northern Syria. The Panamuwa and Bir-Rakib inscriptions (*KAI* 214-215) from Šam'al (Zinčirli) show an overall consistency in both the names of gods and their sequence. The deities named in these inscriptions are the gods of the official pantheon of Šam'al/Ya'udi.[42]

Finally, the term "treaty tradition," as used here, refers to the relatively fixed form in which treaties (and their components) were traditionally cast by a particular treaty-making nation. Thus Hittite treaties and their god-lists followed a standard form for centuries. The same is true of Mesopotamian treaties and their god-lists. The available evidence suggests that these were the dominant treaty patterns in the ancient Near East and that smaller kingdoms which functioned as suzerains in treaties with vassal nations modelled their treaties on the Hittite and/or Mesopotamian traditions. The term "treaty formulary," as used in this study, is functionally synonymous with "treaty tradition."

## Chapter II

## PREVIOUS DISCUSSIONS OF THE LIST OF DIVINE WITNESSES

The list of divine witnesses reads as follows:[1]

> Ἐναντίον Διὸς καὶ Ἥρας καὶ Ἀπόλλωνος, ἐναντίον δαίμονος Καρχηδονίων καὶ Ἡρακλέους καὶ Ἰολάου, ἐναντίον Ἄρεως, Τρίτωνος, Ποσειδῶνος, ἐναντίον θεῶν τῶν συστρατευομένων καὶ Ἡλίου καὶ Σελήνης καὶ Γῆς, ἐναντίον ποταμῶν καὶ δαιμόνων καὶ ὑδάτων, ἐναντίον πάντων θεῶν ὅσοι κατέχουσι Καρχηδόνα, ἐναντίον θεῶν πάντων ὅσοι Μακεδονίαν καὶ τὴν ἄλλην Ἑλλάδα κατέχουσιν, ἐναντίον θεῶν πάντων τῶν κατὰ στρατείαν, ὅσοι τινὲς ἐφεστήκασιν ἐπὶ τοῦδε τοῦ ὅρκου.

One of the earliest commentators on Polybius in modern times was I. I. Reiske, whose notes on the *Histories* were published in 1763.[2] In the list of divine witnesses Reiske discerned an alternating sequence of Greek and Punic deities, so that Philip's representative swore by one group of his gods, then Hannibal by a group of his gods, as follows:

| | Philippus | Hannibal |
|---|---|---|
| 1. | per Iovem, Iunonem et Apollinem | per Genium Carthaginis, Herculem et Iolaum |
| 2. | per Martem, Tritonem, Neptunum | per deos commilitones et Solem et Lunam et Terram |
| 3. | per rivos lacus fontes | per deos agri carthagi-niensis |

4. per deos Macedoniae et Graeciae | (vacat: occupatum enim iam erat)

communiter ambo:
per deos commilitones
qui huic iuriiurando praesunt.[3]

One can already perceive an imbalance in this schema: there is no counterpart on Hannibal's side to "the gods of Macedonia and Greece," by whom Philip's representative swears.[4]

Nevertheless, this interpretation of the list gained wide acceptance. J. G. Schweighäuser echoed this opinion in his 1792 commentary.[5] In the nineteenth century F. C. Movers,[6] F. Münter,[7] R. Pietschmann,[8] O. Meltzer,[9] and others followed Reiske's theory, which by then had become the accepted interpretation. One factor that seemed to confirm this position was the double formula toward the end of the list, invoking all the gods of Carthage and all the gods of Macedonia and Greece; from this it appeared that the gods listed before this were Punic and Greek.[10]

The first to dissent from this understanding of the god-list in H/P was E. Meyer at the end of the nineteenth century[11]; he understood all of the deities in the treaty to be Punic. Meyer identified the first three divine names--Zeus, Hera, and Apollo--as the Punic deities Baᶜal-Šamem, ᶜAštart, and (possibly) Rešep. He did not present his reasons for identifying the names as he did, nor did he articulate the implications; but in light of what we have seen above, if the first triad consists of Punic gods--and the second certainly does--it follows that Reiske's "alternating sequence theory" must be abandoned. In a later article[12] Meyer developed his position a little further, modifying it to some extent (but still maintaining that the deities in question were Punic). But again he supplied no supporting evidence for his conclusions.

Several years after Meyer's first article appeared, H. Winckler also proposed that all the deities in the treaty were Carthaginian.[13] Winckler argued that since (in his view) the treaty was Carthaginian in origin, the gods named must be Carthaginian only. Unfortunately he went further and claimed that the section of the god-list following *pantōn theōn hosoi katechousi Karchēdona* (that is, the section that mentions the gods of Macedonia, etc.) was a later addition to the text. Needless to say, he has not been followed in this unwarranted assumption.

E. Vassel came to a similar conclusion in a series of articles beginning in 1912.[14] Like Winckler, Vassel believed that since Hannibal had just emerged victorious from the battle of Cannae in 216 B.C. and thus held the upper hand, it was the Carthaginian side that initiated the treaty and dictated its terms.[15] As a corollary of this assumption he noted: "It is morally impossible that Hannibal and his senators who assisted him should give to three foreign gods primacy of place over all the gods of their own nation."[16] From this it follows that the first-named gods are Punic; hence again there can be no alternation of Greek and Punic deities.

Thus the alternating sequence theory began to lose ground toward the turn of the century. It received the coup de grâce at the hands of L. F. Benedetto in 1920, in an article that remains even today one of the more perceptive treatments of this section of H/P.[17] Since that time there have been no serious attempts to revive the theory, and no one today denies that we are dealing with only Punic-Phoenician gods in this text.

But now another question arises. Granted that the gods are Punic, are they also the gods of the official Carthaginian pantheon? The majority of commentators assume that this is the case. One significant exception is G. Ch.-Picard, a noted authority on Carthage. He has argued in a number of publications in recent years[18] that the deities named in the list of divine witnesses are not the gods of the official pantheon of Carthage but rather the personal or familial deities of the Barcide family (to which Hannibal belonged).

This view is based largely on the claim that certain important Carthaginian deities are not mentioned in the H/P god-list:

> All the ingenuity of the various interpreters has not succeeded in finding there [in our treaty] some of the deities who enjoyed a certain popularity in the city of Dido: Eshmoun, Astarte, Shadrapa Dionysios, Demeter and Kore, among others. Kronos, who answers to Baᶜal Hammon, does not appear there either. . . .[19]

Not content to note that these Punic gods have not been found in the god-list, Picard goes so far as to claim that the difficulties involved in trying to discern the gods of the Carthaginian pantheon here are "insoluble."[20]

Were Picard correct on this last point, one would almost be forced to conclude that the god-list does not present the official pantheon of Carthage. But the claim that the difficulties alluded to cannot be overcome is based on the erroneous presupposition that there is a one-to-one correspondence of Greek to Punic deities and vice-versa. So, for example, Zeus can only be Ba$^c$al-Šamem;[21] the high deities Ba$^c$al-Hamon, 'Ešmun, and $^c$Aštart cannot be listed here since they correspond (exclusively) to Kronos, Asklepios, and Aphrodite respectively.[22] But this simplistic conception of the correspondence of Greco-Roman and Punic divinities does not bear up under closer scrutiny. It is important to note that during the Greco-Roman period a Semitic deity might be identified with or assimilated to several Greek or Roman gods. Each identification was made on the basis of one point of similarity common to the two deities in question. For example, at Palmyra (Tadmor) the (originally) Babylonian Nabu was identified with two Greek gods, Apollo as well as Hermes.[23] In the Punic world $^c$Aštart (Greek Astarte) was identified at times with Hera/Juno, no doubt because she was the chief goddess in many areas of Syria-Palestine, and at times with Aphrodite/Venus, because her character was that of a goddess of love and fertility.[24] Hence it is not impossible that the first two divine names, Zeus and Hera, are to be taken together, representing the supreme consort pair in the Punic pantheon. As for $^c$Aštart, one cannot exclude the possibility that she stands behind one of the Greek names in the list. Once the presupposition of one-to-one correspondence is set aside, the possibility emerges that the difficulties mentioned above may not be insoluble after all. And once this is acknowledged, one need not resort to a supposed "familial" group of Barcide gods (for which there is no substantial evidence in the first place)[25] to explain the names in the H/P god-list.

Beyond this there are several reasons to expect that in this treaty Hannibal would invoke the official gods of Carthage rather than his personal deities. Polybius tells us that in their treaties with Rome the Carthaginians "swore by their ancestral gods [*tous theous tous patrōous*]."[26] Here "ancestral gods" surely mean the deities venerated generation after generation by their ancestors, undoubtedly including their Tyrian forefathers who founded Carthage. The phrase can hardly refer to each Carthaginian family's personal gods. Given the conservatism among ancient peoples with regard to legal and ritual traditions, it is unlikely that the central act of the Carthaginian treaty tradition--the oath by the ancestral deities--would have been radically altered in favor of an oath by one's family gods. Picard himself, it should be noted, argues that the Barcides' rise to power marked "a return to old national traditions" in Carthage.[27]

Secondly, an international treaty is a public, not a private document. In a pact between Carthage and Macedonia one would expect the state or official deities, not merely personal or familial gods, to be invoked in the oath.[28] To be sure, there is some question as to whether Hannibal exceeded his authority in concluding this treaty with Philip V. Bickerman notes, "In 218, the Carthaginian Senate . . . emphatically stated that pledges given by Punic generals are not binding for Carthage as being made without the consent of the constituted authorities."[29] But in the first place the treaty is concluded (from the Carthaginian side) not only in the name of Hannibal, although his name stands in first place, but in the name of "Mago, Myrcan, Barmocar, and all the other [?] Carthaginian senators who are with him [i.e., Hannibal], and all the Carthaginians serving with him."[30] A.-H. Chroust argues that the three individuals named after Hannibal are "emissaries plenipotentiary" for the Carthaginian government, empowered to ratify the treaty.[31] Bickerman, on the other hand, argues that the three were members of Hannibal's *consilium*, "the higher officers, responsible for the conduct of operations."[32] But Walbank considers this explanation unlikely.[33] If the three men were in fact emissaries plenipotentiary, then the treaty was an official document of state; in such a case it would be reasonable to expect the gods of state to be invoked in the oath. And yet even if it proved to be true that Hannibal was acting independently of the Carthaginian government in this matter, it seems certain from the text itself that he gave the treaty the appearance of a document of state by concluding it in the name of the Carthaginian senate as well as in his own name. If he adhered to the formulary of an official treaty on this particular point, it is reasonable to assume that he would follow the Carthaginian practice of swearing by the "ancestral gods" of Carthage rather than by "familial" deities.

The evidence presented by Picard for interpreting the god-list in H/P as a list of personal gods of the Barcides is, therefore, tenuous. Only further investigation of the divine names can determine whether in fact the major gods of Carthage are listed here; the issue cannot be foreclosed by Picard's premature conclusion that the problem is insoluble.

## Chapter III

## COMPARISON WITH OTHER ANCIENT TREATIES

One of the most serious problems facing the interpreter of H/P is the lack of directly parallel texts. Our task would be considerably simplified if we had even one other Punic treaty to work with. But only a few treaty documents from the West Semitic world survive: the (three) Aramaic treaties from Sefîre (ca. 750 B.C.) and H/P; of these H/P is the only treaty from the Phoenician- or Punic-speaking world known to exist. We must now see whether the Greek treaties, the Aramaic treaties, or the Akkadian and Hittite treaties contain parallels to H/P as regards the *position* of the god-list within the text of the treaty and the *structure* of the list. In order to do this it will be necessary to consider in some detail the god-lists found in these treaties. It is obvious that we cannot undertake a full-scale treatment here; but in the following pages we shall survey at least briefly the material in question.

### A. The Position of the God-Lists

One point of contact between our god-list and those in ancient treaties from the Near East has to do with the relative position within the document. In H/P the list comes toward the beginning of the text, just after a brief introductory section naming the contracting parties.[1]

This relative position within the document does not correspond to what we find in Greek treaties.[2] First of all, a good many Greek treaties contain no god-list at all. A treaty from the ancient Near East, however, would be unthinkable without a specific series of gods as witnesses.

There is not one example from Syria-Palestine, Hatti, or Mesopotamia of a treaty text in which the explicit mention of the gods has been omitted. Treaties from the Greek world that do contain lists of gods usually have them toward the middle or end of the text; sometimes the list is repeated.[3] But in no case does it appear just after a brief introductory section, as in the case of H/P. Moreover, as we have noted, no Greek treaty contains an introductory section similar to that of H/P; the closest parallels come from Near Eastern treaties.[4]

In Near Eastern treaties from the first millennium B.C. the god-list seems to occur regularly in this position. In a number of these the beginning of the obverse has been broken off, so that it is impossible to be certain whether they had a list of gods near the beginning of the text. This is true of the treaty between Marduk-zakir-šumi I of Babylon and Šamši-Adad V of Assyria, ca. 820 B.C. (M/Š); the treaty between Assur-nirari V of Assyria and Mati'ilu of Arpad, ca. 750 B.C. (A/M); and the treaty between Esarhaddon of Assyria and Ba$^{c}$al of Tyre, 677 B.C. (E/B). But aside from these three broken texts, every Near Eastern treaty from this period has a god-list toward the beginning, after the introductory section.

In the first Aramaic treaty from Sefîre (Sf1)[5] a list of divine witnesses appears at the beginning of the text, just after the introductory section in which the contracting parties are named.[6] The small fragment of a treaty involving Sennacherib (Senn)[7] also contains a list of gods on the obverse, although it is difficult to determine precisely how near the list is to the (lost) beginning of the tablet.[8]

The vassal-treaties of Esarhaddon (VTE)[9] are actually copies of one treaty text involving a number of his vassals in 672 B.C. This is the longest and best preserved treaty from the first millennium B.C. discovered to date. It is the only Assyrian text of this kind preserved to such an extent that we can view its entire arrangement, not just bits and pieces. Just after the introductory section (1-12), which is formally quite similar to that in H/P,[10] comes a god-list. It begins with the phrase "in the presence of" (*ina* IGI) as do the Sf1 and H/P god-lists.

*ABL* 1105[11] is a treaty or perhaps a loyalty oath involving Assurbanipal and certain of his vassals or subjects. The very beginning of the text is broken; but the first preserved lines list several gods. This is without doubt the remnant of a god-list, which was most likely preceded by an introductory section. A fragment of Assurbanipal's treaty with the Qedar tribe (A/Q)[12] contains a brief list of gods on the obverse, again apparently toward the beginning of the text.[13]

Of the treaty between Sin-šum-lišir and Nabu-apil-iddina and others (S/N)[14] only the very beginning and end survive. The beginning of the text is identical in format to the beginning of VTE, though the introductory section of the former is somewhat shorter. Immediately after this comes the phrase [*i-n*]*a ma-ḫar* $^{d}$*x*[. . .], "in the presence of (the god) X[. . .]." Given the exact formal correspondence between S/N and VTE up to this point, this is certainly the beginning of a list of divine witnesses: compare *ina* IGI $^{d}$SAG.ME.GAR $^{d}$*Dil-bad*, "in the presence of Jupiter, Venus, etc." in VTE 13.

When we turn to the treaties of the second millennium B.C. we are concerned chiefly with pacts concluded between Hatti and a vassal or, in a few instances, with an equal party. The most commonly attested pattern places the list of divine witnesses toward the middle or end of the treaty. V. Korošec[15] lists this as the fifth element in the typical vassal-treaty; the only exception to this pattern among the treaties he considered was the treaty of Suppiluliuma I of Hatti with Ḫuqqana of Ḫayasa.[16] In this document the list of gods comes toward the beginning, after the first set of treaty-stipulations. But other treaty texts have come to light since the publication of Korošec's study; in some of these the god-list also occurs toward the beginning of the text. Among the more important of these are the treaty between Arnuwanda I and the people of Išmeriga[17] and several treaties with the Kaškeans.[18]

We see then that each of the first-millennium Near Eastern treaties in which the beginning of the text is preserved contains a list of divine witnesses just after the introductory section; this pattern is also followed in some second-millennium treaties from Hatti. It seems clear that the position of the god-list in H/P follows the Near Eastern treaty tradition in this respect rather than the Greek treaty tradition (Appendix I, Chart 1).

## B. The Structure of the God-Lists

### *1. Greek Treaties*

God-lists in Greek treaties are shorter than the list in H/P. They are also quite simple in structure, consisting merely of the names of the various deities by which one or both parties takes the oath. Sometimes each divine name is preceded by the particle *nē* or *nai*, as in *alēthē tauta nē ton* $DN_1$, *nē ton* $DN_2$, etc.:[19] "(I swear that) these things are true by $DN_1$, by $DN_2$, etc." More often the names are preceded by *omnyō*, "I swear," with the divine names in the accusative.[20] There is no parallel to the use of *enantion* in H/P. The oath-deities are not divided into subsections as in H/P but follow one after the other, sometimes joined by *kai*, until the end of the list.

Divinities frequently invoked are Zeus, Ge, Helios, Athena, Ares, Demeter, and Poseidon. Although some of these are also found in the H/P list, the sequence of the gods there does not correspond to that in the Greek treaties. So, for example, Zeus is often named first;[21] but Hera appears only rarely.[22] Helios and Ge are included a good number of times, though never with Selene (as in H/P) and usually just after Zeus.[23] The list often concludes with a "summary phrase," usually *tous (allous) theous pantas kai pasas*, "all the (other) gods and goddesses."[24] The summary phrases in H/P, although they too contain the words *pantōn theōn*, do not have the feminine adjective and also differ in that they mention the countries of the contracting parties ("Carthage . . . Macedonia and the rest of Greece").

### 2. *Treaties from the Semitic World*

It has been noted[25] that the list of gods and curses at the end of M/Š[26] is almost a verbatim duplicate--although somewhat abridged--of the list of gods and curses in the epilogue of the Codex Hammurapi (CH) dating some 800-900 years earlier[27] (Appendix I, Chart 3). The only major difference between the two lists is that in the treaty Marduk and his son Nabu are named at the beginning, whereas they do not appear at all in the CH list. The reason for this change is theological. "Throughout the First Dynasty of Babylon Marduk was an insignificant god,"[28] clearly not the head of the Babylonian pantheon; this position still belonged to Anu. The first evidence of Marduk's supremacy in the divine realm appears in the reign of Nebuchadnezzar (Nabu-kudurru-usur) I, ca. 1100 B.C.[29] Thus between the writing of CH and the treaty under consideration Marduk had been elevated to the highest position in the Babylonian pantheon. It is important to observe that for all the conservatism exhibited by the god-list in M/Š, its slavish adherence to the wording of the god-list in the CH epilogue, the now supreme status of Marduk demanded that his name be written at the head of the god-list; the name of Nabu immediately follows, since he is often intimately associated with his father. Unfortunately the tablet is broken off before the list of divine witnesses ends. This is the only Babylonian treaty that has come down to us.

The same phenomenon, naming the chief deity first in the god-list, appears in all Assyrian royal documents, including monuments, annals, grants, and treaties. The list of deities preserved in A/M[30] is longer than the one in CH and M/Š (Appendix I, Chart 4). The gods are named with their consorts, from Anu through Madanu (7-13); after this the pattern of pairs is continued, though rather artificially, since the pairs of gods named from this point on are not consorts. The entries are not preceded by "in the presence of" as in H/P. Rather after each pair of deities or deity comes the word *tùm-ma-tú-nu* (or the ditto sign), "you are adjured (by). . . ."[31] But in this list Marduk and Nabu (together with their consorts) are listed after the two great triads from the Sumero-Akkadian pantheon, Anu through Adad (7-9). It is important to note that these in turn are preceded by Assur, who thus heads the entire list. The listing of Assyrian deities is concluded (20) by the Sebetti (Akkadian for "the Seven"). The remaining lines (21ff) have unfortunately suffered damage. But the deities listed here are almost certainly "foreign gods,"[32] that is, gods of the vassal party--in this case Mati'ilu of Arpad.[33]

At this point we must underline the importance of noting the function of the Sebetti entry in this and other treaty god-lists. As E. Dhorme once observed, "Often the name of [the] 'Seven' ends an enumeration of divinities. . . ."[34] In fact this holds true in every treaty god-list in which they are mentioned: A/M, Sf1, E/B, VTE. The Sebetti constitute the major caesura in these lists. Specifically, when they are mentioned they are always the last-named Mesopotamian gods. What follows is either a summary phrase or non-Mesopotamian deities.[35] Unfortunately this demarcating function of the Sebetti in treaty god-lists has generally gone unrecognized with the result that the structure of a number of first-millennium lists has been widely misinterpreted.

The fragmentary god-lists in Senn (Appendix I, Chart 7) do not preserve the name of the god in first position, but this is certainly Assur, who is otherwise not mentioned in the two lists.[36] Several of the gods are named with their consorts; it is probable that Assur at the head of the list was named with his consort, Ninlil.[37] The last entry in the list appears to be partially preserved, DINGIR.MEŠ [GAL.MEŠ], "the [great] gods." As we shall see, "the great gods" or "the great gods of heaven and the netherworld" serves as a summary phrase that regularly concludes the god-list as a whole.

Toward the end of E/B[38] is a list of deities consisting of a series of Assyrian gods (of which only a few are preserved) followed by a number of Phoenician deities (Appendix I, Chart 8). A curse follows the name of each individual Assyrian deity and the names of the Phoenician gods in groups of two or three with the exception of the last entry, which lists only one deity (ᶜAštart) followed by a curse. What is preserved of the Assyrian deities, as well as the opening lines of the text, accords almost exactly with the corresponding sections of VTE. In fact the gods and curses in the two documents match so well that "we can even go so far as to suppose that in the treaty of Baᶜal of Tyre, in the gap between the third and fourth column, the same [gods and] curses were listed as in the first curse section of the treaty of Ramataja [ = VTE]."[39] R. Borger has actually restored some damaged lines in the VTE god-list on the basis of E/B.[40] Again the Sebetti are mentioned at the end of the list of Assyrian oath-gods, in the position corresponding to VTE 464. After this are named Bethel and ᶜAnat-Bethel followed by a summary phrase and then six Phoenician deities. Two questions arise with regard to this list: (1) Are Bethel and ᶜAnat-Bethel to be regarded as Phoenician deities (since they follow the Sebetti) or as belonging to the list of the gods of Assyria? (2) Why does the summary phrase occur before the enumeration of gods has been completed, contrary to its use in every other Assyrian royal document? These questions are taken up in Appendix II.

VTE contains not one but four lists of gods (Appendix I, Chart 9). We shall hereafter refer to these as VTE(A), VTE(B), VTE(C), and VTE(D). VTE(A), which comes after the introductory section (1-12), is a list of deities that divides into two parts: (1) a list of heavenly bodies (13-15) and (2) a standard royal Assyrian god-list (16-24). The list of celestial bodies is included because the treaty was apparently concluded at night.[41] Note that each subgroup is prefaced by *ina* IGI, "in the presence of." This god-list is clearly demarcated by ruled lines in the text before and after it. VTE(A) is immediately followed by VTE(B) (25-40), which differs from it slightly. After each entry is the word *ta*[*m-mu*] (or the ditto sign), "(DN) is invoked upon oath."[42] This particular list is characterized by a series of ten lines mentioning "all the gods" of various areas under Assyrian rule. These break down into Assyria (31-37), Babylonia (38-39), the foreign nations (40a), and the concluding summary phrase, "all the gods of heaven and the netherworld" (40b). This list too is set off from the preceding and following sections by ruled lines. VTE(C) differs from both (A) and (B) in several ways. The arrangement of deities is somewhat different. The list is longer than the others (414-93). Each deity's name is followed by one or more epithets and by one or more curses. Each deity, together with the curses with which he or she is associated, is marked off by lines in the text. Note again that the Sebetti are mentioned last (464). After them come several foreign gods (466-71) followed by curses (the text is damaged at this point).[43] Then comes a standard concluding phrase, "the great gods of heaven and the netherworld, who dwell in (all) the world, as many as are named on this tablet" (472-73), to which is joined a long series of curses (474-93). There is also another list VTE(D), beginning at line 518. This is a fourth, distinct series of gods and not originally part of VTE(C), as R. Frankena claims.[44] However, this section "forms no unity."[45] It is extremely disorganized with gods intermingled with simile curses, one god being named several times,[46] and no hint of hierarchical arrangement--contrast (A), (B), and (C)--except that here too Assur is listed in first position (518). This almost haphazard concatenation of gods resembles no royal Assyrian god-list, so that one may doubt if it should be called a god-list at all.

The list of gods preserved on the reverse of *ABL* 1105 is incomplete, being broken off toward the end (Appendix I, Chart 10). This list resembles VTE(C) in a number of respects. Each of the gods is followed by an epithet and curse (or curses). Like VTE(C) each deity with epithet and curse(s) is separated from the preceding and followed by a line. Fourteen divine names are preserved. As usual, Assur is named in first place. After Sin, Ea, and Šamaš is a series of war-gods (Ninurta, Nergal, Zababa, Palil [17-20]); compare A/M with a similar series of war-gods (Ninurta, Uraš, Zababa, Nergal [11-12]) after the two great triads of the Sumero-Akkadian pantheon[47] plus Marduk and Nabu. The position of Marduk and Nabu in *ABL* 1105 is unusual, however. In royal Neo-Assyrian lists these two gods are almost never placed in so high a position, preceding even Šamaš and Sin. This unusual feature may be a sign of deference to the Babylonians, who are probably the party taking this oath.[48] Unique, too, is the fact that the oath is in the first person plural, not the second person plural as in the case of A/M, Senn, E/B, and VTE(C).

A/Q contains two short god-lists (Appendix I, Chart 11). The first is broken, but apparently consisted of no more than three entries. The editors restore the name of Assur in first position; Ninlil (Assur's consort) follows; a third divine name cannot be read with certainty. The first four names in the list on the reverse are the same as in VTE 414-22: Assur (restored), Ninlil, Sin, and Šamaš. These are followed by Marduk (restored) and Nabu in their usual position in Neo-Assyrian god-lists. Ištar of Nineveh and Ištar of Arbela follow, after which the text breaks off.

On the obverse of S/N only traces of the first divine name are preserved (Appendix I, Chart 12). This may well be ᵈS[AG.ME.GAR], that is, Jupiter, as in VTE 13.[49] In this case S/N, which is structurally identical to VTE up to this point, would likewise contain a god-list beginning with celestial bodies. Note here [*i-n*]*a ma-ḫar*, "in the presence of," preceding the divine name, the same phrase as in VTE 13; compare *enantion* in H/P. What is found on the reverse of the tablet is hardly a god-list but rather a section corresponding to VTE 397-401, 472, 475-77. There are several curses invoking the great gods, followed by several curses associated with Sin. The reason for mentioning Sin at this point is probably more literary than theological.[50]

Thus we may discern three sections of the first-millennium Mesopotamian treaty god-list. (1) In first position is the supreme god of the pantheon--in Babylon, Marduk; in Assyria, Assur. (2) Then follows what we might call a "canonical" list of high gods of the Babylonian-Assyrian pantheon. In Assyria this series frequently ends with the Sebetti. In a number of examples several foreign gods are mentioned afterwards. (3) The list as a whole is concluded by a summary phrase. Although the inclusion of such a phrase as a standard feature of the god-list is not obvious, because in so many of the texts the conclusion of the list is missing, it may be safely inferred from (a) the fact that those treaty god-lists of which the ending is preserved do contain such a phrase (Senn [Obv], E/B, VTE[A, B and C]), (b) the fact that other types of official Neo-Assyrian lists of gods regularly conclude with a summary phrase, and (c) the fact that the lists of gods and curses in *kudurrus*--the type of document with the closest parallels to treaty god-lists--almost always conclude in this way (Appendix I, Chart 13).

Sf1 has several points of correspondence to the Mesopotamian god-lists discussed so far. Like VTE(A) and S/N (Obv), the god-list immediately following the introduction begins with *qdm*, "in the presence of" (Appendix I, Chart 5). In fact each series of deities in the list (usually a pair) is preceded by *qdm*. This arrangement finds its closest parallel in H/P, in which each series of deities (usually a triad) is preceded by *enantion*.[51] The meaning of "in the presence of" is the same in Sf1, S/N, VTE(A), and H/P. In each case it refers first of all to the fact that the treaty oath is taken before images of the oath-gods. Bickerman notes that the Greeks too "made solemn statements 'in the presence of,' that is, facing divine images of their own gods."[52] In a Neo-Assyrian letter the writer tells how "surrounded by the images of your gods [*ina lìb-bi* DINGIR.MEŠ-*ka*] I took the oath of loyalty [*a-de-e*] to Your Majesty."[53] But the phrase also had a legal connotation in the Near East.[54] In Mesopotamia the names of witnesses to legal transactions of various kinds are commonly preceded by IGI ( = *maḫar/pān*),[55] "in the presence of." Thus in Sf1, S/N, VTE(A), and H/P this phrase indicates not only that the treaty was concluded in the presence of images of the gods but moreover that these gods were understood to be witnesses to the conclusion of the treaty. In Greek treaties, on the other hand, one does not find *enantion* in connection with the god-lists. Moreover, the notion of the deities as witnesses to the pact is not as clear in Greek as in ancient Near Eastern documents.

The first position in the god-list in Sf1 is held by a pair of deities, of which the first name is missing owing to the state of the text. The second is *mlš*, an otherwise unknown deity.[56] J. A. Fitzmyer is undoubtedly right when he states, "One would expect the first two deities mentioned to be the principal gods of KTK."[57] After this entry, the first two-thirds of the god-list (Marduk through the Sebetti) consists almost exclusively of Babylonian-Assyrian oath-gods, most of whom are also found in the Assyrian god-list in A/M. But whereas A/M lists the two great triads of the Sumero-Akkadian pantheon in exact canonical order, the arrangement in Sf1 seems somewhat haphazard. The first great triad (Anu, Enlil, Ea) is not mentioned at all; however, it is often missing also in official Neo-Assyrian lists. Marduk and Nabu are named just after the first deities. The fact that Erra and Nergal precede members of the second triad (Šamaš, Sin) would not be expected in a Mesopotamian god-list. Hence, as in the case of the first-millennium Mesopotamian treaties, immediately after the supreme gods of the pantheon comes a list of oath-gods from the Mesopotamian treaty tradition. It must be emphasized that the Mesopotamian character of these deities is the essential feature of this section. No doubt these are at the same time the gods of the official pantheon of KTK, although the relationship between KTK and Mesopotamia is not clear.[58]

Even the enigmatic *nkr* (10) belongs to the Mesopotamian treaty tradition. Fitzmyer claims that no attempt to identify this deity or the accompanying *kd'h* has been successful.[59] H. Donner and W. Röllig make no definite proposal for identification,[60] nor does J.C.L. Gibson,[61] who cautiously suggests that *nkr* and *kd'h* were members of the pantheon of KTK. Oddly, only one of these studies--Fitzmyer's[62]--even alludes to M. Jastrow's significant remark on the deity Nin-kar: "Nin-kar is perhaps identical with Nin-karrak, a designation of Gula."[63] Nin-karra(k) and Gula are goddesses of healing. Gula is mentioned toward the end of the E/B treaty (Rev IV.3-4), just before the Sebetti (who terminate this section of the god-list) and in the corresponding place in VTE (461-63). In A/M she is placed beside her consort, Ninurta (Rev VI.11). In a number of Babylonian god-lists Nin-karra(k) is also named near the end. In the canonical god-list from the Ur III period (Appendix I, Chart 2) she is the last deity mentioned by name and is named immediately before the summary statement, "the great gods in their totality," which terminates the list. Likewise in the CH epilogue she is named last, just before the summary statement, "the great gods of heaven and the netherworld." In Sf1 *nkr* is named toward the end of this section of the god-list, just before what appears to be a summary phrase.[64] Given these structural parallels in official Mesopotamian god-lists, there should be no hesitation in identifying *nkr* (i.e., **Nikkar* < **Nin-kar*) with the goddess Nin-karra(k). The identification of *kd'h* is still unsolved; but since this deity is paired with Nikkar, he or she too must be associated with the oath-gods of the Mesopotamian tradition.

The phrase that comes after *nkr wkd'h*, "all the gods of RḤBH and 'DM," has been taken by the majority of commentators as the major dividing line in the god-list.[65] But this cannot be correct. It is true that in Mesopotamian royal inscriptions a summary phrase usually ends a list of gods; however--and this is usually not realized--such phrases do not divide god-lists into sub-sections.[66] Moreover the line in question is not the summary phrase in this god-list. Such phrases, whether in Mesopotamian or Hittite treaties, are of two types: (1) a phrase mentioning the gods of the contracting parties (standard in Hittite treaties), or (2) a phrase mentioning the (great) gods of heaven and the netherworld (standard in Assyrian treaties). The summary phrase in this list is "all the g[ods of KTK and Arpad]," which occurs in its expected position at the end of the list as a whole.[67]

Another reason that commentators see a division at this point is the mention of Hadad of Aleppo, whose name is certainly to be restored after the summary phrase.[68] Perhaps because of the proximity of Aleppo to Arpad, the native land of Mati'ilu (here Mati$^{c}$el), it is usually thought that Hadad of Aleppo begins the list of the gods of the Arpad pantheon. But the inclusion of this deity in a list of oath-gods is part of the age-old treaty tradition of a broad area of the ancient Near East.[69] Hence the naming of this god simply continues the canonical Mesopotamian section. That this is the case is evident also from A/M, where $^{d}$IM $^{uru}$*Ḫal-la-ba*, "Adad of Aleppo" (Rev. VI.18), is followed by more Mesopotamian deities, Palil and the Sebetti; in A/M the section listing non-Mesopotamian gods does not begin until Rev IV.21. Furthermore, to claim that a new section begins after "all the gods of RḤBH and 'DM" necessitates the postulate that the reference to the (Mesopotamian) Sebetti has been inserted where it does not really belong, that is, in a list of Canaanite gods.[70] But once one sees that the principal divider in the list is not this phrase or Hadad of Aleppo but rather the Sebetti (as in other first-millennium treaties), the problem of a supposed insertion disappears. One should note also that with the exception of the supreme gods ([DN] *wmlš*), *nr*, and possibly *kd'h*, all the deities named in Sf1 up to and including the Sebetti are also listed in the contemporary treaty A/M;[71] but none of those listed after the Sebetti in Sf1 is found in A/M. This is another reason for seeing the major break in the god-list after the Sebetti. Hence the entries from Marduk through the Sebetti in Sf1 belong to Mesopotamian tradition; none of the gods of Arpad is mentioned here.

So far it is evident that the structure of the list of divine witnesses in Sf1 follows what we have seen in the first two sections of other first-millennium treaties: (1) the supreme deities of the local pantheon (here [DN] *wmlš*); (2) a (Mesopotamian) list of oath-gods (*mrdk* through *sbt*).

On the analogy of other first-millenium treaties, the first entry of the next section should begin a list of foreign gods. In one sense the next section is that--at least insofar as the gods there are not Mesopotamian. But it is not a list of the gods of Arpad as one might expect.[72]

This final section of divine witnesses opens with *'l w$^{c}$lyn*.[73] It is not clear whether we should read here "El and $^{c}$Elyon" (two gods) or "El-and-$^{c}$Elyon" (one god). Fitzmyer admits the latter possibility, but notes: "In the preceding pairs it was always a question of distinct deities, which suggests that ʾEl and ᶜElyān are distinct Canaanite gods."[74] But first of all, not all of the previous entries are pairs; Hadad of Aleppo (10-11) and the Sebetti (11) are not. The same may also be true of one entry that appears to be a pair--*šmš wnr* (9). The consort of Šamaš (the sun-god) is Aya in Mesopotamia (as in A/M Rev VI.9), whereas here we have *nr* (i.e., **Nūr*, "luminary"). Theoretically *nr* could be an epithet of Aya, but it is never applied to her elsewhere; on the other hand, it is a standard epithet of Šamaš.[75] Thus it would be possible to read *šmš wnr* as "Šamaš-and-Nur" (one deity).[76] Secondly, in the Ugaritic ritual text 608[77] we also have a number of pairs of deities, intermingled with compound divine names--thus *ʾi[l.w]ḥrn* (Obv 13), "ʾEl and Ḥoron," and *yrḫ.w.ršp* (Obv 15), "Yariḫ and Rešep", but *[kṯ]r wḫss* (Obv 18), "Koṯar-and-Ḫasis," a single deity. Hence the fact that most of the preceding entires in Sf1 are true pairs does not mean that *ʾl wᶜlyn* must be so understood. Besides, we have seen that this entry, since it comes after the Sebetti, belongs to a section of the god-list distinct from the one containing the pairs of Mesopotamian oath-gods.

The connection between *ʾl wᶜlyn* and *ʾl wᶜlyn qnh šmym wʾrṣ*, "ʾEl ᶜElyon, creator of heaven and earth," in Gen 14:19 has been pointed out by several authors.[78] What seems not to have been generally noticed, however, is that the order of the three pairs of natural phenomena that immediately follow *ʾl wᶜlyn* corresponds strikingly to those in the beginning of the creation account in Genesis 1:[79]

| Sf1 | Genesis 1 |
|---|---|
| *šmy[n wʾrq]* (11), | *ʾt hšmym wʾt hʾrṣ* (1:1), |
| "Heaven and Earth"; | "heaven and earth"; |
| *[mṣ]lh wmᶜynn* (11-12), | *thwm . . . hmym* (1:2), |
| "Abyss and Springs"; | "deep . . . waters"; |
| *ywm wlylh* (12), | *ywm . . . lylh* (1:5), |
| "Day and Night" | "day . . . night." |

The sequence "heaven and earth" followed immediately by cosmic aqueous elements (the deep, springs, seas, etc.) also occurs in cosmic or cosmogonic contexts elsewhere in the OT and other Near Eastern literature.[80] The fact that *ʾl wᶜlyn* is named just before this abbreviated cosmogony, then, is hardly coincidental. Furthermore, as we shall see, a "creator-god" is sometimes named just before the cosmic elements invoked in the Hittite treaties. All of this suggests an intimate connection (if not identification) of ʾEl-and-ᶜElyon with ʾEl to whom the epithet "creator of the earth" is almost exclusively applied in West Semitic texts.[81] This in turn would seem to argue further for taking *ʾl wᶜlyn* as the name of a single god.

This final section of gods, then, consists of (most probably) one deity followed by three pairs of divine natural elements. Although some of these pairs are occasionally met in Mesopotamian lists,[82] they are never found in Mesopotamian treaties. Rather, they come from the *Hittite* treaty tradition. In fact this entire section of Sf1 (from *ʾl wᶜlyn* to the end of the god-list) derives from that tradition rather than the Mesopotamian. In order to demonstrate this it will be necessary to depart from our intended order of discussion and to mention several relevant aspects of this Hittite tradition.

After the traditional summary statement in the Hittite treaties ("[all] the gods and goddesses of Hatti, [all] the gods and goddesses of GN [the vassal land]"), there comes a final section containing (1) the "olden gods" and (2) divinized natural phenomena. What these two sub-groups have in common is that, in contrast to the deities before the summary phrase, they are not venerated in the Hittite cult and have no temples.[83] We may characterize them, therefore, as the "non-cultic" witnesses.

The first sub-section, that of the "olden gods," consists of a certain group of deities with proper names; they have lost their status in the cult (at least in Hatti) and have been long ago banished to the netherworld.[84] In the Hittite texts they are variously termed:

| Hittite | | Akkadian |
|---|---|---|
| *karūileš šiuneš* | olden/former gods | |
| *uktureš šiuneš* | eternal gods | *ilāni dārûti*, *ilāni ša dārâtim* |
| *kattereš šiuneš* | lower [netherworldly] gods | |
| *taknaš šiuneš* | gods of the netherworld[85] | *ilāni ersetim* |

These are equivalent to the Sumerian Annunaki.[86] For the sake of consistency we shall hereafter refer to them as the "olden gods." They form an extremely stable group within the Hittite treaty tradition.[87]

The second sub-section consists of natural phenomena, again ordinarily in a definite sequence; they include some that are also found in Sf1 (heaven, earth, springs). These, however, are not part of the olden gods, even if they are often associated with them.[88] First of all, in none of the Hittite treaties are these elements ever preceded by the divine determinative, in contradistinction to all other entries in the list, including the olden gods.[89] Secondly, as we noted, the *karūileš šiuneš* are clearly gods of the netherworld; obviously, natural elements such as winds, clouds, etc., do not belong in the netherworld. In several Hittite treaties, moreover, a line is drawn on the tablet separating the olden gods section from the list of natural phenomena;[90] such lines are not arbitrary and almost invariably mark off distinct (sub) sections. Furthermore, in one important early list of gods, in a treaty with the Kaškeans, the phrase *kar-ru-ú-i*[*-l*]*i-aš* DINGIR.MEŠ, "the olden gods," is separated by the summary phrase from four natural elements: heaven, earth, mountains, rivers.[91] Finally, Hittite prayer- and ritual-texts which name the *karūileš šiuneš* never include natural phenomena among them.

It is easy to see how the three pairs of cosmogonic elements in Sf1 correspond to the divinized elements in the Hittite treaties. But what is the connection between *ʾl wᶜlyn* and the olden gods? It cannot be netherworldly character; for while the olden gods are clearly chthonic, there is nothing to lead us to conclude that ʾEl, ᶜElyon, or the combination thereof was ever considered to be such. Nor can ʾEl (and/or) ᶜElyon be considered "non-cultic" like the olden gods.

In order to solve this problem we must first realize that there is nothing known so far in West Semitic religion precisely comparable to the olden gods--that is, a particular group of formerly high gods who have been banished to the netherworld. This is why it is impossible to consider (*ʾl w*)*ᶜlyn* an olden god. As is evident from the inclusion of natural phenomena in the god-list and other features we shall discuss below, the last part of the Sf1 list is patterned on the Hittite model. As regards *ʾl wᶜlyn*, all one need postulate in order to explain the choice of this deity at this place in the list is some point of correspondence between him and some god or gods in the olden gods section of the Hittite treaties. We need not assume that the drafters of this treaty were familiar with all the complexities of Hittite religion and its "thousand gods." Now we have seen that the juxtaposition of *ʾl wᶜlyn* to a series of cosmogonic pairs in Sf1 can hardly be ascribed to coincidence, since ʾEl (or ʾEl ᶜElyon in Gen 14:19) is *the* creator-god in West Semitic literature. At the end of the olden gods section in the Hittite treaties--that is, just before the divinized natural elements--one finds Enlil (with his consort, Ninlil).

Enlil too was associated with creation in Mesopotamian myths; for example, it was he who split heaven and earth with a pickax,[92] whereas the Hittite "Song of Ullikummi" seems to attribute this act to the olden gods.[93] In Hatti Enlil is often equated with Kumarbi;[94] and at Ugarit Enlil, Kumarbi, and 'El are identified in a trilingual god-list.[95] Kumarbi too is sometimes mentioned along with Enlil near the conclusion of the olden gods section.[96] In other words, the god *'l wᶜlyn* is invoked here in Sf1 most probably because he shares one specifying function of the last-named of the olden gods--that of creation.

This analysis of the final section of gods in the Sf1 list leaves no room at all for the gods of Arpad--except, of course, in the summary expression that follows. To some it might occasion surprise that the vassal's gods are completely left out of the list. But this is quite normal in Hittite treaty-writing.[97] In Hittite treaties, if the gods of the vassal party are specifically named at all, it is only in a second, separate list of divine witnesses, which usually follows immediately upon the first.[98] In Sf1, however, we have only one, which corresponds to the first god-list in the Hittite treaties and thus contains no foreign gods (that is, no gods of Arpad).

The final part of the Sf1 list, after the enumeration of the gods themselves, contains a line that functions as both a summary phrase and a statement of divine attestation to the treaty; it is followed by a direct summons to the gods. Although the stela is damaged at this point, the commentators agree that the first line in question is to be read: *šhdyn kl '[lhy ktk w'lhy 'rpd]* (12),[99] "All the g[ods of KTK and Arpad are] witnesses (to this treaty)." This sort of phrase, "(All) the gods of $GN_1$ and $GN_2$," has been encountered in the first-millennium Mesopotamian treaties and is found in the Hittite treaties as well, although not in this position. The first sentence is most probably to be taken as declarative rather than imperative.[100] The next sentence, however, is clearly imperative: "Open your eyes (O gods) to look upon the treaty of Bir-Ga'yah [with Matiᶜel, the king of Arpad!]." Here too we encounter a feature unknown in the Mesopotamian treaties of the period, namely, a direct statement/summons to the gods to act as witnesses to the treaty.[101] But this is a standard feature of the god-lists in the Hittite treaties and is also found at the end of the list of divine witnesses in H/P. In fact, the very wording and sequence--calling the gods to be "witnesses" and to "look upon" the treaty --appear to derive from the Hittite god-lists.[102]

As a result of the foregoing analysis of the structure of the god-list in Sf1 we discern an almost identical structure in this list and the one in H/P[103] (Appendix I, Chart 6). Whereas the number of lines or entries in each section is not the same as in Sf1, there are exactly three lines (each preceded by *enantion*)[104] in each of the three sections of the gods in the H/P list. The one element that does not seem to correspond is *theōn tōn systrateuomenōn*, which has nothing to do with the creator-god ʾEl-and-ʿElyon in Sf1. But the striking overall correspondence to a treaty from another area of the Near East some five hundred years earlier can hardly be coincidental. And as we are about to see, some features of the H/P list of divine witnesses are paralleled in texts a thousand years earlier--the Hittite treaties.

### 3. *Hittite Treaties*

We have already noted the influence of the Hittite treaty-form on one West Semitic treaty; moreover parallels of H/P with the Hittite texts have been recognized by Bickerman. It will therefore be necessary for us to devote particular attention to the structure of the god-lists in the treaties from Hatti.

The Hittite treaties exhibit two types of such lists: a short list just after the introduction to the treaties, and a much longer list, usually after the section listing the treaty stipulations. Since the latter form is normative in later Hittite treaties and is also more frequently encountered, we shall deal first with its structure. Obviously the nature and scope of our study does not permit an exhaustive analysis of these lists.

Given the impact the Hittite treaties have made on ancient Near Eastern studies, particularly in the area of OT covenant research, it is surprising that only recently has a thorough analysis of the list of divine witnesses in these treaties appeared, in an article by G. Kestemont.[105] But Kestemont's study, although an important contribution to our understanding of such documents, is deficient in several respects (see Appendix I, Chart 14). First, the designations he uses for the major divisions of the list are so general as to be all but useless: Section I "consists essentially of masculine divinities";[106] Section II "is formed by the mention of feminine divinities" with "a certain number of masculine divinities";[107] and Section III "is composed of gods of a general character."[108]

It would be more precise to describe Section I as "Principal God(desse)s and Associated Deities." All of the gods in this section are either the supreme deities--the sun-god of heaven, the sun-goddess of Arinna, and her consort, the storm-god--or deities in some way associated with the last. Thus Kestemont's Group B, the "dynastic deities," consists essentially of storm-gods.[109] His Group C lists Šeri and Ḫurri, bull-gods who draw the storm-god's chariot, and the divine mountains Namni and Ḫazzi, mountains sacred to the storm-god.[110] Group D consists of an extensive list of various local storm-gods.[111]

Kestemont's Section II, as described by him, is not only vague in the extreme but inaccurate. The majority of the deities listed here are in fact masculine.[112] The gender of the "protective deities" ($^{d}$LAMMA) in Group A is difficult to determine.[113] But the only clearly feminine deities in this section are Išḫara, Ḫepat, and Ištar[114] in Group C.

Although Kestemont's description of this section leaves much to be desired, the task of finding a more accurate label remains a difficult one. The same holds true for a number of the other subsections of the god-list. In attempting to redivide the list so that the resulting sections (and subsections) correspond more closely to the natures and/or functions of the deities therein, we shall make use of four data: (1) the overall structure of the list, (2) the information we have concerning the nature and function of the deities listed, (3) the dividing lines used by the Hittite scribes themselves in the writing of the god-list, and (4) the short forms of the god-list found in some Hittite treaties.

As in the case of the Semitic treaties we have considered, the god-list in the Hittite treaties appears to follow a generally hierarchical pattern. That this is so can be seen from an overview of its contents. The first section consists of the highest gods, followed by associated deities (Kestemont's I). The list ends with the "non-cultic" witnesses, consisting of the olden gods (Kestemont's III E, which he calls "traditional gods")[115] and what Kestemont correctly designates the "divinized physical elements" (his III F).[116]

The demarcation line between the cultic and non-cultic witnesses seems to be the summary statement, namely, "(all) the gods and goddesses of Hatti, (all) the gods and goddesses of GN [the vassal land]." This phrase appears in all the god-lists (except some of the short ones). The word "all" appears in the majority of the texts,[117] as we might expect in a summary statement; elsewhere in the lists proper "all" occurs with the olden gods[118] or--again in a summary function--with "the gods of the army/camp."[119] In several of the texts the general summary phrase is preceded by a ruled line on the tablet.[120]

dEREŠ.KI.GAL ( = Allatum = Lelwani = "the sun-goddess of the netherworld")[121] is usually named just before or just after the summary phrase.[122] The position after the phrase would seem the normal one,[123] since she is the head of the olden gods,[124] who, as we have seen, are also netherworldly deities. But since she is sometimes mentioned in a "higher" section of the god-list, perhaps her position before the summary phrase is an attempt to place her as close as possible to the olden gods without removing her from the realm of the high gods.

Thus we may consider the summary statement, the listing of the olden gods, and the divine natural elements to constitute the final section of the god-list. At least these labels seem more satisfactory than Kestemont's nebulous "gods of a general character."

We now return to the middle sections of the god-list. The deities listed here (from dLAMMA to the gods just before the summary phrase) are intermediate high gods apparently in descending hierarchy. This intermediate section--or more precisely, its first subsection--is clearly separated from the principal and associated deities (our Section I) by a ruled line in a good number of the treaties.[125]

The first subsection consists of a series of "protective deities." Some of these seem to possess a warlike character, as for example dLAMMA kuš*kursaš*, "LAMMA of the shield," which concludes this subsection.[126] The reading of dLAMMA, however, is notoriously problematic.[127] Kestemont includes in this group Ea/Enki, Allatum/Lelwani, and Damkina; but in all likelihood these go with the following group, namely, the Telepinu cycle. E. von Schuler observes: "Apparently native Hittite theology . . . felt that Telepinu was related to Babylonian Tammuz, since in the lists of divine witnesses he follows after the Babylonian deities Ea and Damkina or the netherworld goddess Allatum (Mesopotamian Ereškigal)";[128] like Tammuz, Telepinu seems to have been a vegetation-deity.[129] In some lists[130] as many as three goddesses appear after Telepinu; their relation to him is in many cases not clear; but that of Nisaba[131] (goddess of grain) to Telepinu is obvious, if in fact he was a vegetation-god. Some of these goddesses may be displaced here (for example, Ištar in W1) from the next section, in which they are usually found.

The next section, which Kestemont describes as "five different sub-groups,"[132] may be more precisely termed "Oath-God(desse)s and Related Deities." The first entry is "the moon-god, lord of the oath" (Akkadian $^{d}$XXX *bēl māmīti*). Then comes Išḫara, who is entitled "queen of the oath" (Akkadian *šarrat māmīti*); sometimes she is placed at the end of this section to form a sort of inclusion: the section thus begins and ends with deities specifically "of the oath."[133] Since Išḫara is the first major goddess to be mentioned after the sun-goddess of Arinna (*pace* Kestemont), the cycle of Ḫebat usually finds its place here; Ḫebat was almost equal in rank to her consort Tešub in the Hurrian pantheon[134] and hence is sometimes listed in Section I (after the other deities associated with the storm-god) rather than here.[135] The mention of Išḫara, who "belongs to the circle of Ištar,"[136] also makes this the logical place to put the Ištar cycle, including her female attendants, Ninatta and Kulitta.[137] In a number of treaty texts this "Oath-God(desse)s" section is marked off from the preceding by a ruled line.[138]

The $^{d}$ZA.BA$_{4}$.BA$_{4}$ section follows; at least in one case it is separated from the preceding by a line.[139] This section consists entirely of war-gods; $^{d}$ZA.BA$_{4}$.BA$_{4}$ is the Sumerian name of the war-deity and in Hittite may have been read Ḫešue.[140] To this section should be assigned the [DIN]GIR.MEŠ KARAŠ *ḫu-u-ma-an-te-eš*, "all the gods of the army/camp,"[141] since the association with war-gods is obvious;[142] these "gods of the army/camp" seems to be mentioned only here.[143] The inclusion of Marduk at this point[144] is also unique and hence no attempt should be made to create a special subgroup for him. Finally Allatum[145]/Lelwani *tâk-na-aš* $^{d}$UTU-*uš*[146] ("Allatum/Lelwani, the sun-goddess of the netherworld") also in Kestemont's subsection II E 1, is simply displaced from our Section III and hence should also not be separately categorized.

The next section is sometimes marked off from what precedes by a line[147] and usually begins either with Apara of Šamuḫa or Ḫantitaššu of Ḫurma. E. Laroche[148] gives no clue as to the nature or function of any of these gods, although Kestemont and Gurney label them "local gods."[149] Each is connected with a cult-place, which thus distinguishes them as a group from the deities in the next section.

We have labeled the last section of the high gods "Lowest-Ranking Gods Venerated in the Cult." It includes several divinized mountains and "the gods (of the?) Lulaḫḫi, the gods (of the?) Ḫapiri/SA.GAZ." Whether these two terms represent deities or nationalities is not clear.[150]

M. Greenberg considers the two groups "legitimate elements of Hittite society."[151] It is clear that the Hapiri often served as contingents in the armies of Hatti[152] and Ugarit.[153] Thus several authors have characterized these deities as gods of the mercenary troops. M. Liverani asserts, "we are dealing [here] with the deities of the mercenary troops, of Iranian and Syrian provenience respectively."[154] This is also the opinion of Gurney, who calls them "the gods of the foreign mercenaries."[155]

Since we have already discussed the "non-cultic" witnesses, we will proceed to the last part of the Hittite god-list. A summons to the abovementioned gods (in the imperative mood) to be witnesses to the treaty and to guarantee its observance occurs either at the conclusion of the god-list[156] or just before the listing of the divine witnesses.[157] Moreover, in most cases where the summons comes at the end of the list a line is drawn just after this[158] indicating that this summons belongs with the god-list.[159] For this reason we have labeled the call to witness Section X. The phrase in its simplest form runs, "May they [i.e., the abovementioned gods] be witnesses at (the conclusion of) this treaty."[160] The gods are also called upon to "stand by"[161] at the treaty (that is, to be present as witnesses), to "listen to"[162] the treaty stipulations (for the purpose of guaranteeing their observance), and (rarely) to "look upon"[163] the treaty.

An important clue to the way the Hittites themselves conceived the divisions of the god-list or rather to what they considered the representative entries in each section may be derived from the earlier short god-lists. Only a few of these are preserved. They differ from the standard (longer) lists not only in length but also in their far greater variation in the order of deities named. It is generally thought that these represent an abbreviated version of the standard god-lists. Laroche notes, for example, "It [the short god-list in KBo VIII 35 Obv II.9-12] condenses the standard lists by keeping only the paragraph headings . . . ."[164] Von Schuler describes such an abbreviated arrangement as *stichwortartig*.[165] But in a recent study Gurney has argued on palaeographical grounds that the short list is "a forerunner, not a condensed version, of the standard lists."[166] We shall consider in detail here only the best preserved and most comprehensive of these short lists, that found in KBo VIII 35 Obv II.9-12:[167]

(9) the sun-god(dess?),[168] the storm-god, ZABABA, LAMMA, Z[itḫariya], (10) Ištar, Išḫara, "lord (!) of the oath," the gods of heaven, (11) the gods of the netherworld, the "olden gods," the gods of Hatti, (12) the gods of Kaška, heaven, earth, mountains, rivers. (13) Let them be witnesses to this treaty.

The first two entries represent our Section I. $^{d}$ZA.$BA_4$.$BA_4$ and $^{d}$LAMMA correspond to Sections V and II respectively.[169] The list also includes another "protective deity," Zitḫariya. The next group, consisting of Ištar and Išḫara, "lord (!) of the oath,"[170] answers to our Section IV.[171] The list then has a summary phrase, "the gods of heaven, the gods of the earth [ = the netherworld]."[172] Perhaps it is because of the mention of the latter that the olden gods (*ka-ru-ú-i*[-*l*]*i-aš* DINGIR.MEŠ) are listed immediately afterwards; one might have expected this phrase to precede the natural elements, inasmuch as these gods are ordinarily listed before them.[173] The summary statement, "the gods of Hatti, the gods of Kaška," corresponds in content and position to what we find in the longer god-lists; it constitutes our Section VIII. The listing of a few divinized elements represents our Section IX. Finally, the call to witness corresponds to Section X.

Thus of the sections we have included in our schema of the standard treaty god-lists, only Sections III, VI, and VII are not represented in this short list.[174] One should also note that like the standard god-lists, the short form does not include foreign gods if there is only one list. In KUB XXIII 77a Rev these are enumerated in a second, separate list.[175]

Having analyzed the basic structure of the list of divine witnesses in the Hittite treaties, we must now compare this structure with that of the god-list in Sf1 and H/P. In the Hittite treaties the order is usually as follows:

1. high gods
2. summary phrases
3. "non-cultic" witnesses (viz., olden gods and divinized natural elements)
4. the summons to witness

Sf1 has approximately the same four major sections, only there the summary phrases come toward the end of the text:[176]

1. high gods
2. gods paralleled in the Hittite treaty tradition (viz., ʾEl-and ᶜElyon [corresponding to Enlil in the olden gods section] and natural elements)
3. summary phrases
4. statement/invocation of divine attestation.

The arrangement in H/P is virtually identical to that in Sf1 (Appendix I, Chart 6):

1. high gods--three lines
2. gods paralleled in the Hittite treaty tradition (viz., gods of the mercenary or ally troops [corresponding to the gods of the Lulaḫḫi and Ḫapiri]--one line, and natural elements--two lines)
3. summary phrases--three lines
4. statement of divine attestation.

As in the case of the sentence after "Day and Night" in Sf1, the *summons to witness* in H/P is declarative rather than imperative. As in the Hittite treaties and Sf1, it marks the very end of the god-list.

The first two of the three *summary phrases*, mentioning the gods of Carthage and Macedonia respectively, reflect in good Semitic idiom[177] precisely what the Hittite treaties read at this point (*mutatis mutandis*), even to the inclusion of the word "all" found in most of the Hittite summary phrases.[178] The reference to "all the gods of the army" in the H/P list also appears to parallel Hittite [DIN]GIR.MEŠ KARAŠ *ḫu-u-ma-an-te-eš*,[179] which serves as a summary phrase at the end of the war-gods section.

We have termed the second part of the H/P god-list *gods paralleled in the Hittite treaty tradition.* The last two lines contain divinized natural elements, answering to our Section IX B of the Hittite lists. The first line, "gods of those who take the field with (us)," that is, of allies or mercenary contingents,[180] corresponds to Section VII B.

Finally, like the Hittite treaties and Sf1, the first section of the god-list in H/P contains the *high gods*. But the correspondence to the Hittite lists may be more detailed. As we have seen, the short god-lists contain the major components of the standard lists. The first three sections are represented by $^{d}$UTU and $^{d}$U/IM, the supreme deities; $^{d}$LAMMA, the "protective deity"; and $^{d}$ZA.BA$_{4}$.BA$_{4}$, the war-god. When one considers how the structure of the rest of the H/P god-list parallels the Hittite list in every line, it may be that the same arrangement of supreme deities, protective deity, and war-deity at the head of the three lines of this section of the Carthaginian treaty is no coincidence:

| Hittite | | H/P |
|---|---|---|
| $^{d}$UTU $^{d}$U | supreme deities | *Zeus, Hēra* |
| $^{d}$LAMMA[181] | protective deity | *daimōn Karchēdoniōn* |
| $^{d}$ZA.BA$_{4}$.BA$_{4}$ | war deity | *Arēs* |

We have seen that in Sf1 the influence of the Mesopotamian tradition dominates to the extent that the high gods section consists exclusively of Mesopotamian oath-deities; and that even the sequence of the high gods reflects aspects of this tradition. The Carthaginian list could hardly be expected to contain Hittite deities; but part of the sequence of the Hittite high gods may be present here, undoubtedly mediated through a long Phoenician treaty tradition of which only this text has chanced to survive.

Chapter IV

# COMMENTARY ON THE LIST OF DIVINE WITNESSES

## A. The High Gods

The task of identifying the nine deities here is not a simple one. One is tempted to find simple solutions and short-cuts. One such solution, as we have seen,[1] would be to proceed as if there existed a set of invariable identifications between Greco-Roman and Punic gods. Since the evidence from the Greco-Roman period reveals that identifications of this kind were more complex, it would be methodologically inadvisable to identify the gods in H/P on the basis of supposedly "standard" assimilations in this period--to maintain, for example, that Zeus must be Ba$^{c}$al-Šamem.

A second simplistic solution, in our opinion, involves a particular use of the "Canaanite pantheon" in reconstructing the pantheons of Tyre, Carthage, etc. To be sure, one may validly speak of a Canaanite pantheon in the sense of the totality of divinities worshipped in Canaan and those areas which adopted the Canaanite religion. But one should exercise great caution in extrapolating from the arrangement of this Canaanite pantheon in one city-state of Palestine to its arrangement in another, especially if a considerable difference in time is involved. Specifically, one may not gratuitously assume that the ranking of the gods and their mutual relationships were identical *in every respect* at all times and in all places. As W. F. Albright has wisely cautioned, "We must always remember that different places and different periods arranged the pantheon differently. . . ."[2] This is not to say that the Canaanite pantheon was radically rearranged from place to place and period to period; rather the evidence shows a great deal of stability in terms of hierarchy and consort relationships among the gods. For example, at the head of this pantheon in all areas always stands one of the three highest gods: ʾEl, Dagan, or Hadad.

In the wake of the discoveries at Ugarit, which have enormously amplified our understanding of the Canaanite pantheon, one is tempted to fill in gaps in our knowledge of the pantheon in other areas of Syria-Palestine from the data gathered from this important site. As A. Caquot has recently noted:

> the discovery of a genuine corpus of "Canaanite" myths [at Ugarit] immediately led to a temptation that must be resisted: to believe that Ugarit provides testimony for *the* West Semitic religion in the Bronze Age . . . and that religious vestiges scattered elsewhere must be organized around what the Ugaritic literature has shown us.[3]

Succumbing to this temptation leads to the view that, for example, since ʾEl was the head of the pantheon at Ugarit he must be the supreme deity in every site where the Canaanite pantheon is venerated. This view is clearly reflected in a statement by R. A. Oden, Jr.: "Hadad is not the head of the Canaanite/Phoenician pantheon; that position is held firmly by ʾĒl."[4] In the same article Oden goes on to claim: "there is no compelling evidence to assert that Dagan was the head of the Canaanite/Phoenician pantheon."[5]

But the evidence in fact does not support Oden's position. It is at least possible that Dagan was the chief deity of the pantheon venerated by the Philistines according to the books of Judges and 1 Samuel--a possibility that even Oden cannot deny.[6] It is even more likely that Dagan held this position in the pantheon of Iron Age Arpad, judging from A/M. If E. F. Weidner's reading is correct, the first-named and thus highest-ranking of the foreign--that is, Arpadian--gods in that treaty (Rev IV.21) is $^{d}$[*D*]*a-g*[*an ša*]$^{u}$[$^{ru}$G]N, "[D]ag[an of] (the ci[ty) G]N."[7] And our present knowledge of the pantheon emerging from Ebla indicates that Dagan was the chief deity there too.[8]

As for Hadad, note the supreme position of Ba$^{c}$al-Šamem ( = Hadad)[9] as seen from his first-ranking position in royal inscriptions from tenth-century B.C. Byblos (*KAI* 4.3), ninth/eighth-century B.C. Hamath (*KAI* 202 B.23), and eighth-century B.C. Karatepe (*KAI* 26 A III.18-19). As we shall see, Oden's attempt to prove that Ba$^{c}$al-Šamem is 'El must be judged unsuccessful.[10] And despite the arguments of F. M. Cross and C. W. L'Heureux , the first-ranking position of Hadad (immediately followed by 'El) in several ninth/eighth-century B.C. royal inscriptions from Šam'al (*KAI* 24.15-16; 214.2,11,18; 215.22) indicates that he was the supreme deity of the pantheon in this part of northern Syria[11] as he was at other sites in this area such as Aleppo and Alalakh.[12]

Secondly, one may not validly presume that the consort relationship between various Canaanite deities always and everywhere followed the pattern at Late Bronze Ugarit. Cross betrays this presupposition when he says: "On a priori grounds we should expect Punic 'Ēl to have as his consort 'Ēlat. At Ugarit and in Sakkunyaton 'Ēlat-Asherah is the primary wife of 'Ēl. . . ."[13] Cross notes that at Ugarit $^{c}$Anat and $^{c}$Aštart were also wives of 'Ēl, though 'Ašerah was the chief wife.[14] But on what grounds may one assume that at all times and places where 'El was venerated 'Ašerah was considered his chief consort? And despite Cross' statement, it does not seem clear that 'Ašerah was the primary wife of 'El in Sakkunyaton; this position may have been held by $^{c}$Aštart.[15] As regards the "Punic 'Ēl," we shall see that his chief consort was most probably not 'Ašerah.[16]

### 1. Ἐναντίον Διὸς καὶ Ἥρας καὶ Ἀπόλλωνος

In the survey of the god-list structure of ancient Near Eastern treaties we saw that one consistent feature is a generally hierarchical arrangement of the deities listed therein. More specifically, it was seen that, without exception, the god-list begins with the supreme god or gods of the pantheon in question. That this is also the case in the H/P list is further indicated by the choice of the names Zeus and Hera, the supreme consort pair in the Greek pantheon. Hence there can be little doubt that behind these Greek names stands the supreme consort pair of the Carthaginian pantheon at the time of Hannibal.

*a. Zeus*. Who was the chief god of the Carthaginian pantheon in 215 B.C.? Fortunately, the possibilities are limited to two, Ba$^{c}$al-Šamem and Ba$^{c}$al-Ḥamon. It is certain that both were important deities at Carthage. Ba$^{c}$al-Šamem seems to have been the supreme deity for the Seleucids in Syria, in particular for Antiochus IV Epiphanes, during the second century B.C., under the name Zeus Olympios;[17] for the Syriac text of 2 Macc 6:2 translates the Greek name as *b$^{c}$lšmyn* (the Aramaized form of Ba$^{c}$al-Šamem).[18] Earlier he is listed at the head of a group of deities in several royal inscriptions from Syria-Palestine. The inscription of Yeḥimilk of Byblos, from about the middle of the tenth century B.C. (*KAI* 4), lists *b$^{c}$l šmm.wb$^{c}$l<t?> gbl.wmpḥrt.ʾl gbl qdšm* (3-5), "Ba$^{c}$al-Šamem, and the Lady [?] of Byblos, and the assembly of the holy gods of Byblos." The Karatepe inscription, from about 720 B.C. (*KAI* 26), likewise lists Ba$^{c}$al-Šamem first in a series of gods: *b$^{c}$l šmm wʾl qn ʾrṣ wšmš $^{c}$lm wkl dr bn ʾlm* (A III.18-19), "Ba$^{c}$al-Šamem, and ʾEl, creator-of-the-earth, and eternal Sun, and the whole assembly of the gods."[19] In both of these texts the position of Ba$^{c}$al-Šamem would seem to indicate the supremacy of this god in the two areas,[20] especially if one considers the official character of the documents in question.

Yet this evidence does not justify the position of J. Teixidor, as reflected in the title of a subsection in his recent book on religion in the Greco-Roman Near East--"Baal Shamin, the Chief God of the Phoenicians."[21] The supremacy of Ba$^{c}$al-Šamem in some areas of Syria-Palestine does not necessarily mean that this god was supreme everywhere in this region--for example, at Carthage or Tyre. His rank in a given city can be determined only by an examination of the evidence from that city itself.

We do not possess a great deal of direct testimony about Ba$^{c}$al-Šamem at Carthage. From several inscriptions it is evident that he was venerated there. One broken inscription mentions *ḥnʾ khn lb$^{c}$lšmm* (*CIS* 1.379), "-ḥana, priest of Ba$^{c}$al-Šamem." Another reads, *qbr ḥmlkt khn b$^{c}$lšmm*,[22] "The grave of Ḥamilkat, priest of Ba$^{c}$al-Šamem."

Perhaps the most important text from Carthage concerning Ba$^{c}$al-Šamem is a stela dating from the third century B.C. (*KAI* 78). The dedication reads *lʾdn lb$^{c}$l šmm wlrbt ltnt pn b$^{c}$l wlʾdn lb$^{c}$l ḥmn wlʾdn lb$^{c}$l mgnm* (2-4), "To Lord Ba$^{c}$al-Šamem, and to Lady Tanit, 'Face-of-Ba$^{c}$al,' and to Lord Ba$^{c}$al-Ḥamon, and to Lord Ba$^{c}$al-MGNM." This text is unique among the votive inscriptions discovered at Carthage to date in that it places Ba$^{c}$al-Šamem ahead of Tanit and Ba$^{c}$al-Ḥamon. Here Ba$^{c}$al-Šamem appears, it is true, to have preeminence over the other divinities.[23] But one must exercise caution in drawing conclusions from such a document.

J. G. Février has used this text to explain the first four deities named in H/P; thus Zeus = Ba$^{c}$al Samem, Hera = Tanit, Apollo = Ba$^{c}$al Ḥamon, and the *daimōn* of the Carthaginians = Ba$^{c}$al-MGNM, whom Février identifies with Ešmun.[24] Aside from a number of other difficulties this interpretation encounters, one should bear in mind that the stela is a private dedicatory inscription of a private individual, whereas H/P is an official document of international import--one, moreover, in which the list of deities shows influence from the Near Eastern treaty tradition. We should expect no such influence, however, in a text such as we are dealing with here. The supplicant who set up the stela was at liberty to dedicate it to what deities he pleased. Mesopotamian hymns and prayers often address a particular god in the most flattering terms, going so far as to assert that "there exists no other god than the one being addressed."[25] In such cases, of course, the supplicant is not concerned with the "official" rank of the deity within the pantheon. Similarly the listing of Ba$^{c}$al-Šamem ahead of the other deities in *KAI* 78 cannot be taken as a reflection of the official standing of this god in the Carthaginian pantheon but is at most an indication of the worshipper's desire to pay special honor to him for some reason or other. This fact, plus the fact that the position accorded to Ba$^{c}$al-Šamem at Carthage is unique to this inscription, makes it unwise to draw any conclusions from *KAI* 78 about his rank in the official pantheon.

The popularity of Ba$^{c}$al-Ḥamon at Carthage is abundantly attested from votive inscriptions. By far the majority of these begin: *lrbt ltnt pn b$^{c}$l wl'dn lb$^{c}$l ḥmn*, "To Lady Tanit, 'Face-of-Ba$^{c}$al,' and to Lord Ba$^{c}$al-Ḥamon." It is evidently for this reason that several scholars, notably S. Gsell, hold the opinion that "Tanit and Baal Hammon were quite probably the principal gods of the city [Carthage]. . . ."[26] But here, too, one should be cautious. A god's popularity, reflected in such phenomena as votive stelas, is not an infallible index of his hierarchical status, as Février has aptly observed with regard to Ba$^{c}$al-Ḥamon.[27]

Thus far the evidence from Carthage does not allow us to decide which was the supreme deity. We must therefore turn to Tyre. The founders of Carthage brought their religious beliefs to the African continent from the Phoenician mother-city and continued to maintain close religious as well as economic and political ties with the homeland. It is axiomatic that outposts such as Carthage, remote from the change and development of the main cultural centers, tend to exhibit a high degree of conservatism and to preserve aspects of the mother culture that gradually disappear elsewhere.[28] This conservatism would be all the more expected in the religious realm; and in fact Carthage does reveal precisely this kind of conservatism.[29] Hence if we can determine which of the two deities --Ba$^{c}$al-Šamem or Ba$^{c}$al-Ḥamon--was the chief god of the Tyrian pantheon, we might well be in a position to answer the same question with regard to the pantheon of Carthage.

In light of the preceding remarks about the conservatism of Carthage vis-à-vis the religious traditions of the mother-city, it should come as no surprise that Ba$^{c}$al-Šamem and Ba$^{c}$al-Ḥamon were major deities at Tyre as well as Carthage. There is little in the way of epigraphy from Tyre that is helpful here. A votive stela from Umm el-$^{c}$Awāmīd (between Tyre and Acco), which dates from 132 B.C. (*KAI* 18), mentions Ba$^{c}$al-Šamem but no other gods. More significant is the report from Josephus that Hiram of Tyre, an ally of Solomon, built a causeway to "the temple of the Olympian Zeus" (*tou Olympiou Dios to hieron*) and "adorned it with offerings of gold."[30] We have seen that in other sources this Olympian Zeus is identified as Ba$^{c}$al-Šamem,[31] and there is little doubt that this is the god in question here as well. As for Ba$^{c}$al-Ḥamon, Diodorus Siculus[32] informs us that Kronos ( = Ba$^{c}$al-Ḥamon)[33] was one of the "ancestral gods" (*theoi patrōoi*) of the Carthaginians--hence undoubtedly a god brought to Carthage from Tyre. The fact that the citizens of Carthage were accustomed to offering their noblest sons to Kronos (a practice that no doubt had its origins on the Phoenician mainland) is an indication of the importance of this god at Carthage (and Tyre). But which god was the *chief* deity of Tyre?

To answer this question we turn to a document of singular importance for this phase of our study, the treaty between Esarhaddon and Ba$^{c}$al of Tyre (Appendix I, Chart 8). Fortunately the text preserves intact a list of Phoenician deities invoked as witnesses to the treaty.

We must note several things about these deities. They are explicitly called in the summary phrase "the gods of Trans-Euphrates [ = Palestine]" (DINGIR.MEŠ *e-bir* ÍD); thus they are clearly non-Assyrian deities. But they are not merely gods of Phoenicia in general--although most if not all of them may have been widely venerated there. In the context of such a treaty they can only be the gods of the vassal party (in this case, Tyre) following immediately upon the state gods of Assyria. Hence J. T.

Milik is wrong in his supposition that these deities "are evidently 'transfluvian' [i.e., "Trans-Euphratean"] but not necessarily Tyrian. . . ."[34] There would be no point in enumerating a random series of Phoenician gods unless all were important to Tyre. Thus there can be no doubt that the gods invoked here are specifically Tyrian deities, by which Baᶜal swears allegiance to Esarhaddon.[35]

Secondly, we must ask whether one can discern anything about the arrangement of deities in the list. On that point Février remarks that "it does not seem that the deities are enumerated according to a strictly hierarchical order."[36] This statement may be partially correct. But a study of Assyria's dealings with foreign nations shows that the listing of foreign deities in official documents was not a casual affair.

Assyrian royal inscriptions of the Sargonid period consistently name the supreme god(s) of a foreign pantheon first in lists of foreign deities. In recounting his sack of Susa, Esarhaddon mentions the gods of Elam, which he carried off as booty to Assyria. He names nineteen gods, the first of which is (In-)Šušinak.[37] Like Marduk in the Babylonian pantheon,[38] In-Šušinak had risen from a relatively minor position in the Elamite pantheon as city-god of Susa (his name means "Lord of Susa")[39] to become "the great national god."[40] In-Šušinak's rise to supremacy in the pantheon had taken place by the reign of Šilḫak-In-Šušinak[41] toward the end of the second millennium. Esarhaddon, narrating his restoration of the gods of Der to that city, lists first "the Great Anu (and) the Queen of Der."[42] These are the supreme deities of Der, god and consort.[43] In fact in every list of the gods of Der in Assyrian royal inscriptions "the Great Anu" [ = Ištaran/Sataran][44] is named first.[45] In another inscription Esarhaddon lists the Arabian gods he returned to Hazail, king of the Arabs. First in the list of the six deities is ᶜAtaršamain.[46] It is all but certain that she was the chief deity of these tribes.[47]

The listing of the supreme deities of foreign pantheons first was no accident or mere act of diplomatic courtesy. The passages cited above give lists of gods deported to Assyria. The Assyrians had a political purpose in carrying off the images of these gods intact, as D. J. Wiseman explains with regard to the gods of Arabia: "Sargon [II], Sennacherib and Esarhaddon had all used their capture of the gods of the Arabs to bargain for more effective control over the desert tribesmen who constantly harassed the western Assyrian provinces."[48] Given the fact that the more important a particular deity was to a nation (i.e., the higher it ranked in the official pantheon) the more bargaining power the Assyrian king would hold by its capture, the Assyrians were selective in their despoliation of foreign gods:

> The evidence points to a policy of selective capture of [divine] statues. It seems clear that many small shrines and their images were irreverently destroyed, while other religious objects were spared and taken off to Assyria. Presumably the treatment of each god and statue accorded with the importance attached to them by the Assyrian conqueror and his advisors; those items most revered by the vanquished nation were exiled.[49]

Sargon II, having entered Muṣaṣir, the capital city of Urartu, proceeded at once to give orders that (the image of) Ḫaldia, "the chief deity of Urartu," be removed from the city.[50] He deported not only Ḫaldia but his consort, Bagbartu, as well.[51] This incident is a clear illustration of the strategic importance from the Assyrian point of view of knowing the chief gods of a particular area. From these considerations there can be little doubt that the Assyrians named the chief god of Tyre in first place in the list of Tyrian gods in E/B.

Now the question is, which divine name begins the list of Tyrian deities? A number of writers, among them W. F. Albright,[52] D. J. McCarthy,[53] and J. Teixidor,[54] believe that this is Ba$^{c}$al-Šamem. This view is based on the assumption that the foreign (i.e., non-Assyrian) gods section of E/B begins after the summary phrase. Hence these authors do not consider $^{d}$*Ba-a-a-ti*-DINGIR.MEŠ $^{d}$*A-na-ti-Ba-a*[*-a-ti*-DING]IR.MEŠ[55] ("Bethel [and] $^{c}$Anat-Bethel"), who are named just before this phrase, to be Phoenician deities. If this view were correct, that is, if the summary phrase were the dividing line in the list, then Ba$^{c}$al-Šamem would be the first-named foreign god. In this case Oden would be fully justified in his claim that in E/B "Ba$^{c}$al Šamēm is placed at the head of the pantheon. . . ."[56]

But none of the above-named commentators on this document presents any evidence for his division of the list with the summary phrase as the caesura. In the absence of any documentation for this conclusion one must regard it as based on nothing more than a cursory examination of the list. A careful analysis of the structure, however, leads inescapably to the conclusion that it is the Sebetti, not the summary phrase, that divides the list. Because the presentation of arguments for this point would constitute an unduly long digression here, the evidence supporting the contention that the Sebetti is the dividing line and that Bethel and ᶜAnat-Bethel are Tyrian deities will be presented in Appendix II. Suffice it to say at this point that in his important article on the vassal treaties of Esarhaddon Frankena also takes the position that Bethel and ᶜAnat-Bethel are the first-named foreign deities in this list,[57] as does Korošec in a recent study of deities invoked in cuneiform treaties.[58] Since Bethel and ᶜAnat-Bethel are named first among the gods of Tyre, there can be no question that the Assyrians understood them to be the supreme deities of Tyre. And since, as we have seen, knowledge of which deities were supreme in a foreign pantheon was a matter of strategic importance to the Assyrians, there is no reason to believe that the drafters of this treaty were mistaken in this case.

Given that Bethel and ᶜAnat-Bethel must be the supreme gods of the official Tyrian pantheon, how does one explain the fact that the writing of the name "Bethel" here does not reflect the Phoenician pronunciation of this divine name? There are several apparent peculiarities in the name $^{d}$*Ba-a-a-ti*-DINGIR.MEŠ. It is generally conceded that this represents in Akkadian script the theophorous element which in West Semitic is **bêt-'il* (*bêt-'ēl* in Hebrew);[59] the name means "House/Temple of El/God." But the first element of the Akkadian form in E/B is to be read *bayt*. In Phoenician the word for "house" would be **bêt* (like the construct in Hebrew), with the contracted diphthong.[60] Then why does the Akkadian text not read *$^{d}$*Be-e-ti*-DINGIR.MEŠ? This peculiarity cannot be explained as an "Assyrianism," since in the Assyrian dialect of Akkadian "house" is likewise *bēt*(*u*).[61] The form *bayt* here can be explained only on the basis of Aramaic[62] (or some similar language or dialect) in which the diphthong /ay/ was generally uncontracted at least until the eighth century B.C.[63] The word "house" in Aramaic was pronounced /bayt/, not /bēt/, even when the word occurred in the construct.[64]

The second element of the name in E/B is DINGIR.MEŠ, which strictly speaking corresponds to *ilī/ilāni*, "gods"; whereas the Phoenician (underlying the divine name *Baitylos* or *Bētylos*[65] in Sakkunyaton)[66] as well as the Aramaic (*byt'l*)[67] presume *'l*, "El/God," the singular form. This apparent anomaly can be explained from Akkadian scribal convention in writing West Semitic names:

> There is no orthographic regularity in the representation of the West Semitic theophorous element *bêt'ēl*; it occurs twice as É.DINGIR.MEŠ and twice as É.DINGIR. Nevertheless . . . it is apparent that the normal way of writing *'ēl* was DINGIR.MEŠ.[68]

There is no doubt that the plural form DINGIR.MEŠ corresponds to the singular *'il/'ēl:*

> The identification of the theophorous element DINGIR.MEŠ with *'ēl* is certain, for we have Aramaic endorsements in which two of the names [written in Akkadian] . . . are also given in Aramaic script. Thus . . . the same name [is] written in cuneiform as ^I^*ra-ḫi-im*-DINGIR.MEŠ and alphabetically as *rḥm'l*; similarly, the name ^I^*ḫa-za-'*-DINGIR.MEŠ . . . is written in the endorsement of the same tablet *hzh'l*.[69]

The Akkadian form ^d^*Ba-a-a-ti*-DINGIR.MEŠ, then, was read *Bayt-'il/'ēl* (i.e., "Bethel"), not **Bayt-ilī/ilāni*. This being so, there is no connection between the divine name here (note the DINGIR determinative) and the *byty 'lhy'*, "betyls" (note the plural form of the *nomen rectum*) in the second treaty from Sefîre.[70]

Hence the form of the name that appears in E/B evidently has a Mesopotamian and Aramaic history. Albright points out that "the name [Bethel] did not begin to appear as an element in theophorous names until cir. 600 B.C."[71] This fact would indicate that the deity had been introduced into Mesopotamia by Aramaic-speaking inhabitants or immigrants sometime prior to this date--thus around the time of the writing of E/B or (more likely) somewhat earlier. But the Aramaic form of the name does not necessarily mean that Bethel was (originally) an Aramaic deity as Albright claims.[72] Rather it may be explained from the fact that Bethel was already (i.e., by the middle of the seventh century B.C.) a deity venerated in Mesopotamia by the Arameans and hence known to the Assyrians; for by this period Mesopotamia had become highly Aramaized. The Assyrian drafters of the treaty wrote the name in the Aramaized form familiar to them, ᵈ*Ba-a-a-ti*-DINGIR.MEŠ.

There is really no reason to deny that Bethel and ᶜAnat-Bethel were Phoenician deities. The name Bethel probably does not occur at Ugarit,[73] but Sakkunyaton lists *Baitylos* (or *Bētylos*?)--distinct from *baitylia*, "betyls"--as a god.[74] R. Zadok has recently expressed the opinion that the name É.DINGIR-*a-di-ir* (Bethel-addir or perhaps -ādir), the name of a West Semite in sixth-century B.C. Mesopotamia, may in fact be Phoenician.[75] The goddess ᶜAnat, of course, is well known from Ugaritic literature and is attested in a few Punic names.[76] The form ᶜAnat-Bethel turns up elsewhere only in later Aramaic documents;[77] but compound names of this type are well attested, particularly in the Phoenician-Punic world.[78]

A study of the Phoenician god-list from E/B makes it quite clear that Baᶜal-Šamem was not the head of the Tyrian pantheon in the seventh century B.C. But who is the god Bethel? Can he be identified with any better-known Canaanite or Phoenician deity? And what is his relation, if any, to Baᶜal-Ḥamon?

From the phrase *bt'il* at Ugarit (even though this was most probably not a divine name in the Late Bronze Age) and from Sakkunyaton's juxtaposition of 'El (*Ēlos*) and Bethel (*Baitylos* [or *Bētylos*?]),[79] we can be relatively certain that the name should be translated "Temple of 'El" rather than "Temple of God."[80] The use of the term to denote a "betyl," that is, a sacred stone which any deity might inhabit, seems to represent "an independent, parallel development."[81] Therefore we postulate that Bethel was (originally) *an hypostasis of 'El*. E. Kraeling has recently expressed a similar view:

> The close association of Baityl with 'El [in Sakkunyaton] suggests that the two names are artificially differentiated. Since the fact that Bethel means "house of 'El" is indisputable, we may suppose that it is simply a personification of 'El's house (in heaven) and as such was a substitute expression for 'El.[82]

Significant too in this regard is the fact that

> *El* is completely absent from the Elephantine onomasticon. This is most noteworthy since the other elements had by then all but disappeared from Israelite names in Babylonia and Judah as well but *El* was found in about twenty-six names from Judah and about twenty names from Babylonia.[83]

It would thus seem that at Elephantine too the name Bethel had supplanted that of ʾEl, as took place at Tyre (according to our hypothesis) at least two centuries earlier.

Like *šmwʾl* ("the name of ʾEl") and *pnwʾl* ("the face of ʾEl") in OT names, the Phoenician **btʾl* would also be what we might call a "cultic hypostasis" of ʾEl, "referring to manifestations of ʾĒl available to the worshipper."[84] This tendency toward hypostatization of the deity is well known not only in Israelite religion[85] but in Canaanite-Phoenician religion as well.[86] One might compare, for example Ugaritic (*ᶜṯtrt*) *šm bᶜl* (*CTA* 16.6[127].56), "(ᶜAštart [of?]) the name of Baᶜal" (cf. *KAI* 14. 18. *ᶜštrt šm bᶜl*) and Punic (*tnt*) *pn bᶜl*, "(Tanit [of?]) the face of Baᶜal."[87] Nor does one have to theorize that Bethel originally denoted the actual temple of ʾEl (on earth, that is), which later became deified. In Mesopotamia even where temple names are used as quasi-theophorous elements they are never divinized.[88] Thus we might theorize three conceptual stages: (1) the god ʾEl (as in the Ugaritic texts); (2) Bethel, an hypostasis or "veritable doublet"[89] of ʾEl (as in E/B); (3) Bethel, distinct from ʾEl but still closely related to him as his "brother" (as in Sakkunyaton).[90]

Only if Bethel is an hypostasis of or "circumlocution of ʾEl"[91] could one explain the supreme position of this deity (and his consort) in the pantheon of Tyre according to E/B. Going back to the "pantheon list(s)" from Ugarit,[92] we find that ʾEl either is named first or is listed immediately after the god *ʾilʾib* (which seems to mean "the divine father").[93] After *ʾilʾib* and *ʾil* come *dgn* (Dagan) and *bᶜl* (Hadad), who thus constitute the highest echelon of the Ugaritic pantheon. This high rank of ʾEl, Dagan, and Hadad is also reflected in Sakkunyaton, who has Dagan (*Dagōn*) as the "brother" of ʾEl, and who lists him just after ʾEl and Bethel.[94] As for the rank of Hadad (*Adōdos*), Sakkunyaton specifically calls him "the king of the gods" (*basileus theōn*).[95] The fact that Bethel is named in this highest echelon of gods by Sakkunyaton and is intimately associated with ʾEl (as his "brother") may also point to the original identification of the two divine names.[96] Finally, ᶜAnat-Bethel almost certainly means "ᶜAnat(, the consort) of Bethel."[97] We know from Ugarit that ᶜAnat was one of the consorts of ʾEl (although Ašerah was his chief wife)[98] as well as of Baᶜal. This too seems to point to the identification of Bethel and ʾEl.

But is there a relation between Tyrian Bethel ( = ʾEl) and Carthaginian Baᶜal-Ḥamon? The answer to this question is affirmative, since it is virtually certain that Baᶜal-Ḥamon is in fact ʾEl. Thirty years ago B. Landsberger concluded from certain inscriptions from Šamʾal (Zinčirli) that *bᶜl ḥmn* (*KAI* 24.15-16) is equivalent to *ʾl* (*KAI* 214.11,18; 215:22).[99] Recently Cross has confirmed this hypothesis and pointed out further evidence. He compares Diodorus Siculus' reference to the myth of Kronos' ( = Baᶜal-Ḥamon's) sacrifice of his own children[100] with the myth preserved in Sakkunyaton of ʾEl's sacrifice of Yadīd and Môt, his offspring.[101] Yet more significant is Sakkunyaton's repeated equation of Kronos with *Ēlos* (ʾEl).[102] This would seem to be sufficient grounds to confirm the identity of Baᶜal-Ḥamon and ʾEl, aside from the other evidence Cross presents.[103]

Thus the Tyrian Bethel, an hypostasis of ʾEl and the supreme god of the Tyrian pantheon, is the predecessor of the Carthaginian Baᶜal-Ḥamon, an epithet of ʾEl. Furthermore, we have already seen from Diodorus Siculus that Kronos ( = Baᶜal-Ḥamon) was a god brought by the founders of Carthage from Tyre and that he was of sufficient status that the Carthaginians used to sacrifice their noblest sons to him. Obviously, then, he was no minor deity at Carthage or Tyre. Yet if he is not identical with Bethel, one is hard pressed to account for his absence from the E/B list of major Tyrian deities. Hence we conclude that Baᶜal-Ḥamon was the chief god of Carthage.

R. A. Oden, Jr. has recently attempted to show that Baᶜal-Šamem was identical to ʾEl.[104] His thesis--if correct--would pose problems for our identification of Bethel with ʾEl, since Baᶜal-Šamem occurs with Bethel in the E/B god-list. As we would not expect ʾEl to be named twice in this list, it is likely that one of these proposals is wrong. If Oden is correct, it would mean that at Tyre Bethel was already a god distinct from ʾEl (who would according to Oden be Baᶜal-Šamem in E/B). This in turn would make it unlikely that Baᶜal-Ḥamon would correspond to Bethel at Tyre; rather he would correspond to Baᶜal-Šamem.

Oden uses a variety of arguments in defense of his hypothesis. We should first recall, however, that underlying his argumentation is the presupposition that the hierarchy of the Canaanite pantheon at Late Bronze Ugarit--in particular, the supreme status of ʾEl--was normative for all subsequent periods and locales where this pantheon was venerated.[105] The arguments presented for the identification of Baᶜal-Šamem with ʾEl may be summarized in three areas: epithets, iconography, and function.

As regards *epithets*, one should note first that all the evidence that Oden adduces is, with a single exception, drawn from late sources (i.e., dating from the Common Era).[106] In an Aramaic inscription on a boundary stone from Gözneh (*KAI* 259), dating from the fifth/fourth century B.C., Baᶜal-Šamem bears the epithet *rbʾ*, "the great (one)." From this Oden concludes that Baᶜal-Šamem is supreme among the deities listed.[107] This is correct, but not because of the epithet "great." Following this logic one could argue that in Egypt Rešep enjoyed supreme status and in fact was Baᶜal-Šamem, since he is called *ršp nṯr ᶜ3 nb pt*, "Rešep, the great god, the lord of heaven."[108] Of course, this is not the case. Baᶜal-Šamem is the highest of the three deities listed not because of *rbʾ* but because he is named first among the deities. But this does not suffice to equate him with ʾEl, since ʾEl was not the head of the Canaanite pantheon at all times and places.[109]

The strongest evidence adduced in terms of epithets comes from an Aramaic inscription from Hatra, first/second century A.D. (*KAI* 244), where we read: *bᶜšmyn qnh dy rᶜʾ*, "Baᶜa<l>-Šamem, creator of the earth." Despite J. Teixidor's hesitations,[110] the epithet is almost certainly to be translated "creator of the earth," an Aramaization of the ancient epithet of the creator-god ʾEl.[111] One can only agree with Oden that in *this inscription* Baᶜal-Šamem is equated with ʾEl. But can one argue from a Mesopotamian text so far removed from first-millennium Syria-Palestine for the identification of Baᶜal-Šamem in that region during that period? It is quite possible that this represents nothing more than a late syncretism of Baᶜal-Šamem and ʾEl. No firm conclusion can be drawn from this inscription about the identity of Baᶜal-Šamem, especially if one considers that this deity is given this epithet only here.

Hence Oden's arguments from epithets are far from convincing. A stronger case, in fact, could be made for the identification of Baᶜal-Šamem and Hadad by appealing to epithets from this late period. An epithet applied several times to Baᶜal-Šamem during this period is *keraunios*, "of the thunder-bolt."[112] The conclusion is inescapable that at least one of the features of Baᶜal-Šamem during this syncretistic period points to the traditional picture of Hadad, the storm-god, so often depicted wielding a thunder-bolt.[113] This epithet, common to Baᶜal-Šamem and Zeus, may help to explain why Baᶜal-Šamem is so frequently equated with Zeus during the Greco-Roman period.

Arguments from *iconography* are even less impressive. All of these date from the Common Era. Oden concedes, "The Palmyra representations of *bᶜlšmn* bearing a bunch of grapes or grain could readily portray the god Dagan from whose very name was drawn a word for grain."[114] Again no distinct image of Baᶜal-Šamem emerges from these late, possibly syncretistic portrayals. The solution to the identity of this god must be sought in an earlier period when syncretism was not so pronounced.

Before considering Oden's arguments from the functions attributed to Baᶜal-Šamem, we should note several inscriptions--dating from the pre-Christian period--which contain the name of Baᶜal-Šamem together with that of ʾEl. A third-century B.C. votive inscription from Carthage (*KAI* 78) is dedicated "To Lord Baᶜal-Šamem, and to Lady Tanit, 'Face-of-Baᶜal,' and to Lord Baᶜal-Ḥamon."[115] Oden accepts with Cross the identification of Baᶜal-Ḥamon with ʾEl;[116] hence he is forced to remark, "the collocation of *bᶜlšmn* with *bᶜlḥmn* is less comfortable. . . ."[117] "Less comfortable" is an understatement indeed. First of all, the names are clearly separated. But even if they were juxtaposed they could hardly be equated. There is not a single example from first-millennium god-lists of an outright equation of two deities in juxtaposition; nor is there a single example of the use of the "epexegetical *waw*" in these lists.

If the occurrence of these two names in the same inscription is troublesome for Oden's hypothesis, the incontrovertible juxtaposition of the names Baᶜal-Šamem and ʾEl in the eighth century B.C. Karatepe inscription (*KAI* 26) deals a fatal blow to the identification of the two gods. In this text we read, "Baᶜal-Šamem, and ʾEl, creator-of-the-earth, and eternal Sun [Šamaš], and the whole assembly of the gods."[118] Oden attempts to avoid the obvious conclusion that here Baᶜal-Šamem and ʾEl are two distinct deities--with ʾEl in second place--by resorting to an incredible hypothesis: it is "quite possible," he says, that the three divine names are all appellations of ʾEl![119] Though deities occasionally bear epithets in official inscriptions--as ʾEl and Šamaš do here--the tendency in all such documents is toward succinctness in the use of them. In no West Semitic royal inscription does one ever find such an unwieldly concatenation of names and epithets referring to the same god as Oden would have us read here. And as we noted, proper names of deities are never "equated" in first-millennium god-lists. Comparison of this text with a contemporary royal inscription from nearby Šamʾal (*KAI* 214) shows a similar arrangement of divine names; but here the preposition *l-* before each name demonstrates beyond cavil that each name represents only one deity:

*lhdd. wlʾl.*  *wlrkbʾl.wlšmš*

Compare the Karatepe text (A III.18-19):

*bᶜlšmm wʾl qn ʾrṣ*  *wšmš ᶜlm . . .*

Hence first-millennium B.C. inscriptions from Syria-Palestine list Baᶜal-Šamem and ʾEl as distinct deities. Note, however, that Baᶜal-Šamem and *Hadad* are never named in the same first-millennium B.C. god-list.

We shall now consider Oden's evidence for the Baᶜal-Šamem = ʾEl hypothesis drawn from the *function* of Baᶜal-Šamem. He speaks several times of "function," which in his view is important for identifying Baᶜal-Šamem.[120] This he summarizes as "that particular concern shown by ʾĒl for his worshippers, particularly kings";[121] Oden finds this same "function" predicated of Baᶜal-Šamem[122] and hence infers the identity of the two gods.[123]

The problem with this argument lies in the nature of this function --if indeed it may be so termed: there is nothing specific about it. The mere fact that a king erects a stela to a god who has granted him some favor does not mean that only this god shows a "particular concern" for him. Every deity who answers the prayers of a petitioner "functions" in this way, whether the petitioner be king or commoner. And there are numerous inscriptions from the West Semitic world that testify to the fact that more than one god was accustomed to answering the prayers of those who sought his help.

Clearer evidence of a function specific to Baᶜal-Šamem may be found in a first-millennium B.C. document originating from Palestine. Sakkunyaton makes the following statement with regard to this god: "But when the droughts came, they [i.e., Genos and Genea] extended their hands to heaven, to the sun; for this god alone they considered to be the Lord of Heaven, calling him Baᶜal-Šamem [*Beelsamēn*]. . . ."[124] Although the phrase "to the sun" (*pros ton hēlion*) has been taken as evidence of the "solar character" of Baᶜal-Šamem,[125] most probably it simply reflects that later syncretism of various deities with the sun-god attested during the Hellenistic period and later.[126] Indeed even Baᶜal-Ṣaphon, who was originally identical with Hadad, the storm-god, came to be regarded as a solar deity under the name Zeus Kasios.[127] Oden himself admits, "the Sun is an odd deity to entreat during a drought. . . ."[128] Quite so; yet for obvious reasons he does not press this statement to its logical conclusion, namely that the appropriate god to invoke under such circumstances would be the *storm-god*. In a classic passage from Ugaritic literature (*CTA* 19.1.44-45), it is clear that Baᶜal (Hadad) is responsible for there being "no dew, no rain, no welling-up [?] of the deep. . . ."[129] Since Baᶜal/Hadad had brought on the drought mentioned in this Ugaritic text, presumably one would have been well advised to pray to him to end it. The same logic applies to the Sakkunyaton passage. Thus this passage is best understood to imply that Baᶜal-Šamem was a storm-god rather than the patriarchal deity ʾEl.

The most important single bit of evidence for the function of Baᶜal-Šamem comes from eighth-century B.C. Palestine--the god-list in E/B. In Assyrian treaty god-lists the curses are more often than not tailored to fit the nature or function of the deity with whom they are associated. D. R. Hillers notes "the theological precision with which the writers in many cases assign to a god an epithet and curse in keeping with his nature and function."[130] For example, a curse associated with Ištar, goddess of war, threatens the oath-breaker with military defeat (VTE 453-54); the curse connected with Gula, goddess of healing, expresses the wish that disease be inflicted on the violator of the treaty (VTE 461-63). Even in the case of non-Assyrian deities this practice was frequently followed. In E/B the curse that accompanies the name of ᶜAštart corresponds partly to that of her Mesopotamian counterpart Ištar (E/B Rev IV.18a = VTE 453) and partly to that of Venus (E/B Rev IV.18b-19 = VTE 430), no doubt because both ᶜAštart and Ištar were closely associated with this planet.[131]

Of particular interest is the curse following Ba$^{c}$al-Ṣaphon. The god is asked to "raise an evil wind against your [i.e., Tyrian] ships" (IM *lem-nu ina* $^{giš}$MÁ.MEŠ-*ku-nu lu-šat-ba*)[132] with the expected disastrous consequences for the Tyrian ships. This curse follows *three* divine names, those of Ba$^{c}$al-Šamem, Ba$^{c}$al-Malage, and Ba$^{c}$al-Ṣaphon. The verb *lušatbâ*, though clearly governed by the three deities, is in the singular; only once in this curse does a verb occur in the plural with these gods as subject.[133] What is the significance of the fact that the same curse is associated with three divine names and that at least one verb of which they are the subject is singular? It does not mean, as Teixidor believes, that the three are "equated, if not identified."[134] And despite the fact that all three names appear to be--or originally to have been--epithets, we do not have a case here of a divine name followed by several epithets. Each of the names is preceded by the divine determinative, which incontrovertibly shows that they are three distinct gods. Rather in this text the three gods function together as a unit. A verb in the singular (*liškun*) is also used with the Sebetti in this god-list,[135] which certainly does not mean these seven gods were thought to be one; elsewhere too they govern singular verbs, which indicates that they "act as a unit."[136] Hence in this curse Ba$^{c}$al-Šamem, Ba$^{c}$al-Malage, and Ba$^{c}$al-Ṣaphon function together without losing their respective identities. And yet the very fact that this particular curse, to "raise an evil wind," is attached to all three of them indicates that the characteristic *function* of each has to do with the sea and/or storm. It is particularly appropriate to Ba$^{c}$al-Ṣaphon, whom Albright has styled "the marine storm-god *par excellence*."[137] Ba$^{c}$al-Malage is still a rather elusive figure; but as we shall see,[138] he may be Kušor, a Canaanite god associated with the sea at Ugarit and in Sakkunyaton. The epithet "lord of the heavens" is obviously appropriate to a deity whose specific concern of function relates to the celestial realm, but hardly to a sea-god. It is significant that in Mesopotamia the epithet *bēl šamê* or *bēl šamê u erṣeti* (Sumerian EN.AN.KI.BI.DA), "lord of the heavens (and the earth)," is predicated only of the high celestial gods (Anu, Enlil, Marduk, Šamaš, Sin/Nannar, Adad).[139] In the Hittite treaties the storm-god ($^{d}$IM or $^{d}$U) frequently bears this epithet;[140] in these lists he is always named just after the supreme solar deities, just as here Ba$^{c}$al-Šamem is named after the supreme consort pair, Bethel and $^{c}$Anat-Bethel. Hence it is difficult to avoid the conclusion that in E/B Ba$^{c}$al-Šamem is regarded as a storm-god.

Finally, evidence for the equivalence of Ba$^c$al-Šamem and Hadad may be found in a comparison of the gods listed in the royal inscriptions from Karatepe (*KAI* 26) and nearby Šam'al (*KAI* 214, 215). B. Landsberger thought it probable that *b$^c$l ḥmn* and *'l* were one and the same deity because these two divine names occur at precisely the same place within the list of gods in the royal inscriptions from Šam'al.[141] Similarly in the Šam'al and Karatepe inscriptions the gods *b$^c$l šmm* and *hdd* occur first in the listing of the gods (the numbers below refer to *KAI* 26 A III.18-19; 214.18; 215.22):

| | | | | |
|---|---|---|---|---|
| 26: | *b$^c$l šmn w'l qn 'rṣ* | | *wšmš $^c$lm wkl dr bn 'lm* | |
| 214: | *lhdd.wl'l* | *wlrkb'l.* | *wlšmš* | |
| 215: | *hdd. w'l* | *wrkb'l.b$^c$l.bt.wšmš* | | *wkl.'lhy.y'dy* |

Note that the order of the gods--Hadad/Ba$^c$al-Šamem, 'El, Šamaš--is the same in each case, except that at Šam'al the dynastic god *rkb'l* comes after Hadad and 'El. The Karatepe and Šam'al inscriptions not only come from the same geographical area but from the same time period as well: *KAI* 26 is dated ca. 720 B.C.; *KAI* 214, ca. mid-eighth century B.C.; and *KAI* 215, between 733/32 and 727 B.C.[142]

We conclude that Ba$^c$al-Šamem was not identified with 'El in first-millennium B.C. Syria-Palestine. Rather he was a god "who during the first millennium B.C. had been conceived as a weather god only."[143] In all likelihood he was in fact Hadad, "lord of heaven" being originally an epithet of this deity. Thus we may reaffirm our hypothesis that according to E/B it is Bethel, not Ba$^c$al-Šamem, who was identified with 'El at Tyre.

If we are correct in concluding that Ba$^c$al-Ḥamon was the head of the Carthaginian pantheon, we would seem to be in a position to go further and assert that he is in fact the "Zeus" named in H/P. An objection that arises at this point is that the usual Greek equivalent for Ba$^c$al-Ḥamon is Kronos, whereas Zeus is generally equated with Ba$^c$al-Šamem.[144] We have seen that Sakkunyaton several times identifies (the Phoenician) 'El with Kronos;[145] but he also explicitly equates Ba$^c$al-Šamem with Zeus: *Beelsamēn . . . ho esti . . . Zeus . . . par' Hellēsin*,[146] "Ba$^c$al-Šamem . . . that is, . . . Zeus . . . among the Greeks." And as we noted earlier, Zeus Olympios of 2 Maccabees corresponds to Syriac *b$^c$lšmyn*. Still later Ba$^c$al-Šamem in Syria is consistently equated with Zeus.[147]

But as we have indicated,[148] the identification of a Punic with a Greco-Roman deity was based on a particular point of similarity common to the two deities. So also Baᶜal-Ḥamon/ʾEl was identified with Kronos no doubt primarily because their roles in the theogonies of Greece and Phoenicia were virtually identical: each is said to have castrated his father, "Heaven," and ruled in his stead.[149] But if, as we have argued, Baᶜal-Ḥamon/ʾEl was also the chief god of Carthage, it is quite possible that in virtue of his position as supreme god he might be identified with Zeus on occasion.[150] S. Moscati also argues for "a certain fluidity in the Greek adaptations [to Phoenician deities]: thus, for example, Zeus [in H/P] could be Baal Hammon instead of the usual Cronos. . . ."[151] E. Merkel notes, "the name of the highest Greek god [Zeus] was used not only to designate the great god of heaven [Baᶜal-Šamem] but also the local *baᶜals* who reigned over a [particular] place or tribe."[152] Moreover Baᶜal-Ḥamon was apparently later known as Zeus Ammōn.[153] Thus one should not decide a priori that Zeus must always stand for Baᶜal-Šamem in the Punic world.

As there can be little doubt--judging from the unvaried pattern of all Near Eastern treaty god-lists, in which the supreme deities of the official pantheon are named first[154]--that in H/P "Zeus and Hera" stand for the supreme consort pair within the pantheon of Carthage, we must conclude that Zeus here is Baᶜal-Ḥamon. The translators of the treaty could hardly have used the name Kronos for Baᶜal-Ḥamon. The collocation "Kronos and Hera" would surely have sounded strange to the Greeks: Kronos was not the head of the Greek pantheon nor was Hera his consort.

Finally, Février raises another objection to the equation Zeus = Baᶜal-Ḥamon, viz., that Baᶜal-Ḥamon could not have preceded the mention of Tanit at the time of H/P (in which he believes Hera is Tanit), since "almost all the Carthaginian dedicatory inscriptions mention Tanit before Baᶜal-Ḥamon."[155] Hence he concludes that Zeus here must be Baᶜal-Šamem.[156] His argumentation reveals the common methodological error of failing to distinguish between god-lists from popular religion and those found in official documents.[157] The fact that Tanit precedes Baᶜal-Ḥamon in votive inscriptions may well indicate that her popularity exceeded that of her consort, but it does not necessarily mean that she outranked him in the official pantheon. Moreover, the male consort is usually named before the goddess in treaty god-lists from the Semitic world.[158] Specifically, the fact that this is the order of the supreme deities in E/B (Bethel and ᶜAnat-Bethel) is quite significant and may have been determinative of the order in subsequent god-lists in the Tyrian treaty tradition.

*b. Hēra.* Two Punic deities were traditionally identified with Hera (or Juno): Tanit and ᶜAštart,[159] the two great goddesses known at Carthage and throughout the Punic world. In this god-list, then, Hera would be one of these two. But the choice is already decided if the Zeus of our treaty is Baᶜal-Ḥamon, for it is beyond question that the spouse of Baᶜal-Ḥamon was Tanit. Hence "Zeus and Hera" represent Baᶜal-Ḥamon and Tanit.[160]

Yet it remains for us to discuss the connection of Hera/Tanit in H/P with the earlier treaty tradition. If Baᶜal-Ḥamon, the chief god of Carthage, answers to Bethel at Tyre, does Tanit correspond to ᶜAnat-Bethel?

It is possible that Tanit corresponds to the major Canaanite goddess ᶜAnat. First of all, as Tanit was the spouse of Baᶜal-Ḥamon (the "Carthaginian ʾEl"), so ᶜAnat-Bethel was the consort of Bethel (the "Tyrian ʾEl") just as ᶜAnat was one of the consorts of the Ugaritic ʾEl. The fact that in Sakkunyaton ᶜAnat (*Athēnē*)[161] is represented as the daughter of ʾEl[162] does not necessarily contradict her status as wife of ʾEl; for at Ugarit too she appears to be ʾEl's daughter[163] as well as his consort--not an unusual relationship for Canaanite deities.[164]

Secondly, both Tanit and ᶜAnat show a particularly close relationship to the goddess ᶜAštart. At Ugarit these two goddesses were intimately associated,[165] a phenomenon that led ultimately to Atargatis (i.e., ᶜAštart-ᶜAnat), the great Syrian goddess. One finds them in parallelism in several Ugaritic texts:

*dk nᶜm.ᶜnt.nᶜmh*
*km.tsm.ᶜṯtrt.tsmh*[166]

Whose charm is like the charm of ᶜAnat,
Whose beauty is like the beauty of ᶜAštart.[167]

This parallelism appears to occur also in *CTA* 2.1[137].40:

*[ymnh.ᶜn]t.tʾuḫd.*
*šmʾalh. tʾuḫd.ᶜṯtrt*

[ᶜAna]t seized [his right hand,]
ᶜAštart seized his left hand.

The names of the two deities are juxtaposed in a number of other texts from Ugarit[168] and even in Egyptian documents.[169] Most significantly, an ivory plaque discovered recently at the site of ancient Sarepta is dedicated "to Tanit-ᶜAštart" (*ltntᶜštrt*).[170]

The relationship between Tanit and ᶜAštart was so close that one frequently reads the claim that Tanit is merely the African form of ᶜAštart.[171] This contention is occasionally bolstered by appealing to the fact that the name of Tanit--although she was without doubt an extremely popular goddess--very rarely occurs in theophorous Punic names, whereas ᶜAštart is one of the most common such elements.[172] But the figure of Tanit may well be concealed in the theophorous element *mlkt*, "queen," which is quite frequent in Punic names.[173] In any case one should bear in mind that the name of Baᶜal-Ḥamon, the chief god of Carthage and also the most popular (to judge from the votive inscriptions), occurs rather infrequently as a theophorous element in Punic names: the element *ḥmn* is attested only several times,[174] and even *ʾl* is surprisingly infrequent in these names.[175] Nor is there any evidence of proper names in the vicinity of Tyre containing the element Bethel (or ᶜAnat-Bethel),[176] despite the fact that these were the chief deities of Tyre;[177] these elements are first found in West Semitic names in later Aramaic texts.[178] Therefore one may not infer from the Punic onomasticon that Tanit is identical with ᶜAštart.[179]

Some scholars believe that the inscriptions found at the temple of Hera/Juno at Tas Silg in Malta are evidence of the identity of Tanit and ᶜAštart.[180] Dedicatory texts found there read *lᶜštrt* ("to ᶜAštart") in Punic script and *lt*, *tn*, or *t* ( = *ltnt*, "to Tanit"?) in Neo-Punic script.[181] The fact that the inscriptions to Tanit are consistently written in a later script has been taken by some, therefore, to indicate that here the cult of Tanit supplanted that of ᶜAštart, or rather that the cult of ᶜAštart became the cult of Tanit.[182] This would be evidence, of course, of an identification of the goddesses. But Cross is surely correct when he concludes, "we must construe the mixture of dedications to *ᶜAštart* and to *Tannit* as evidence that the temple originally was dedicated to both. . . ."[183] He reaches this conclusion apparently on the basis of *KAI* 81, which begins: *lrbt lᶜštrt wltnt blbnn*, "To the Ladies, to ᶜAštart and to Tanit-in-Lebanon." The inscription goes on to speak of the erection of two new sanctuaries (*mqdšm ḥdšm*) to the goddesses. Cross remarks, "There is not the slightest reason to doubt the identity of *tnt pn bᶜl* and *tnt blbnn.*"[184] The most significant aspect of this text for our study is that it is from Carthage and dates to the third/second century B.C.[185] This would place it very near the time of H/P itself. It is difficult to deny, then, that in the Carthage of Hannibal's day Tanit and ᶜAštart, though closely associated, were distinct goddesses.[186] Thus the relation between Tanit and ᶜAštart parallels that between ᶜAnat and ᶜAštart at an earlier period.

W. F. Albright holds that Tanit was "the Carthaginian appellation of the goddess Anath. . . ."[187] He notes parallels between a frequent Roman epithet of Tanit, *Virgo Caelestis*, and ᶜAnat's "standing appellation at Ugarit," *btlt ᶜnt* ("Virgin ᶜAnat").[188] *Caelestis* corresponds to another epithet of ᶜAnat, *bᶜlt.šmm.rmm* (602.7), "lady of the high heavens."[189] In Egypt too "lady of (the) heaven(s)" (*nb.t pt*) was applied to ᶜAnat.[190] At Elephantine "queen of the heavens" (*mlkt šmyn*) was applied to a goddess mentioned in the same line with Bethel and thus presumably ᶜAnat(-Bethel).[191] Most recently F. O. Hvidberg-Hansen has identified Tanit with ᶜAnat.[192]

Cross, on the other hand, has argued that it is ʾAšerah, the principal goddess of the Ugaritic pantheon and the chief consort of ʾEl, who most likely stands behind the figure of Tanit at Carthage.[193] While it is entirely possible that certain traits of ʾAšerah were inherited by Tanit, we have found that Tanit corresponds *principally* to ᶜAnat, not ʾAšerah. It will therefore be necessary to consider the evidence Cross adduces for this contention.

We have already noted that a priori considerations play at least some part in Cross' position; but on methodological grounds we cannot accept this kind of argument.[194] Following Albright, Cross interprets the name of ʾAšerah in Ugaritic, *ʾaṯrt ym*, as "she who treads on the sea"[195] or more precisely, "she who treads on the Sea(-dragon)."[196] This derivation is quite plausible but not certain.[197] The name Tanit--or more accurately, Tennit--is then derived from Canaanite **tannin*, "(sea-) dragon," plus the feminine ending *-tu*: **Tennit <*Tannit <*Tannittu <*Tannintu.*[198] According to this etymology the name would mean "the one of the serpent."[199] Cross sees a parallel between this and his translation of *ʾaṯrt ym.*[200] But to accept this parallel one would have to postulate a rather bizarre semantic shift, so that "the One of the [Sea-] serpent" actually means the one who *conquers* the Sea-serpent. Yet it is difficult to see how **tannintu* could mean anything except "(female sea-)

serpent." Cross makes no attempt to account for this semantic shift. He appeals, however, to the epithet *d̲t bt̲n* found in the Proto-Sinaitic texts; he believes the word *tnt* in these texts refers to Tanit,[201] although *tnt* never occurs in the inscriptions which mention *d̲t bt̲n*. Following Albright,[202] he interprets the latter epithet as "the one [feminine] of the serpent."[203] But once again the etymology is not certain; *d̲t bt̲n* could just as easily mean "the one [feminine] of Bašan,"[204] since the relative particle frequently forms divine epithets with place names in Semitic.[205] In any case there is no clearly established connection between *d̲t bt̲n* and Tanit. Cross also argues that the Canaanite epithet *lbʾt*, which he translates "the Lion Lady"[206] (or "the Lioness"), points to Tanit; he calls this "the old epithet of Ašerah."[207] But as we shall see,[208] there is no evidence that ʾAšerah ever bore such a title; and the contention that Tanit was represented with leonine features is not well founded. We must conclude that Cross's arguments for ʾAšerah as *the* predecessor of Tanit are not persuasive.

The treaty tradition of Carthage, as manifested in H/P, seems so far to follow closely what we have seen of the Tyrian god-list. Specifically, there is enough evidence to make it plausible that Tanit's predecessor in the Tyrian treaty tradition was ᶜAnat-Bethel, who was likewise the consort of an ʾEl-deity.

*c. Apollōn.* This seems to be one of the easier identifications in the god-list. The equivalence of Apollo and Rešep is well attested. For example, several fourth-century B.C. bi- or trilingual inscriptions from Cyprus explicitly identify Rešep with Apollo.[209] It is no wonder, then, that the great majority of commentators equate the Apollo of our treaty with Rešep.[210] On the other hand, other suggestions such as an identification with Baᶜal-Ḥamon have proven unconvincing. For one thing, if Baᶜal-Ḥamon is identified with Zeus here he obviously cannot stand behind the name of Apollo. For as H. Otten has aptly remarked, "It corresponds to the nature of these lists of divine witnesses that each deity be named only once."[211] Nor can a connection between the element *ḥmn* (<*ḥmm*, "to be hot") and Apollo as a sun-god be substantiated.[212] The connections between Apollo and Rešep are far more obvious.[213] Rešep is connected with plague at Ugarit (*CTA* 14[KRT].1.18) and in the OT (e.g., Deut 32:23-24; Hab 3:5; Ps 78:49), as was his Mesopotamian counterpart, Nergal.[214] Apollo was also associated with plague, as is obvious from the beginning of the *Iliad* (1.44-52). Moreover, the image of Apollo as the archer par excellence, *Apollōn hekēbolos*, "Apollo the Far-shooter," is paralleled by *ršp ḥṣ* (*KAI* 32.3,4), "Rešep of the arrow," in Phoenician and *b^c^l ḥẓ ršp* (*UT* 1001.1.3), "Rešep, lord of the arrow," at Ugarit.[215]

It is known from a Carthaginian inscription that Rešep was venerated at Carthage and had a temple there; one text (*CIS* 1.251) mentions *^c^bd bt ʾrš[p]*, "the servant of the temple of Aršu[p]" ( = Rešep). This temple is probably the same as that of Apollo near the agora in Carthage noted by Appian.[216] Valerius Maximus records that the image of Apollo in this temple was gilded and housed in a chapel sumptuously plated with gold.[217] This would indicate that Apollo's (Rešep's) temple was a "very important"[218] one at Carthage.

There is no direct inscriptional evidence of a temple of Rešep at Tyre, but the presence of his important sanctuary at Carthage makes it quite likely that he was venerated in the mother-city as well. Diodorus Siculus[219] writes that during Alexander's siege of Tyre a rumor was started that Apollo was about to leave the city. The worried citizens tried to prevent this by tying the statue of the god to its base with golden chains. The same incident is related by Quintus Curtius.[220] It is most likely that this Apollo is actually the Phoenician Rešep.

But the presence of Rešep in H/P does present something of a problem from the standpoint of the treaty tradition. Since it is true that Rešep is listed immediately after the supreme deities of Carthage here and that he was brought to Carthage from the mother-city, how does one explain the fact that he is not mentioned in E/B?

Judging from the position in which the name occurs one might think that Apollo/Rešep presents a "patron" or "dynastic" deity here. We saw above that in the Hittite treaties the mention of a patron deity comes after the naming of the supreme gods. The same is true in a number of Phoenician inscriptions from Šam'al (Zinčirli) in which the god *rkbʾl*, who is called *b^c^l bt*, "lord of the house (i.e., of the dynasty)" and is thus the patron god of the present dynasty, is named just after *hdd* and *ʾl* (*KAI* 214.11,18; 215.22).[221] But if there is no real evidence for Heracles/Melqart as a dynastic deity (as Picard argues), there is even less for Rešep as a patron deity of the Barcides.

But the presence of Rešep here might be explained along other lines. We have seen that in some of the Near Eastern treaties pairs of deities are named in the god-lists; this is true, for example, of A/M and Sfl, which has the closest structural parallels to H/P. The pairs begin with god and consort (A/M Rev VI.7-12; Sf1 A.7-9)--in other words, with natural pairs of deities. Even after these consort pairs the pairing is continued somewhat artificially with other gods whose mutual relationship is not that of consorts (A/M Rev VI.13-15; Sf1 A.11: *ʾl wᶜlyn*). In H/P, of course, the divine witnesses are not listed in pairs but in triads. The question then arises as to whether at least some of these triads consist of deities who are related one to the other by familial ties or whether the pattern of threes is purely artificial. There is no doubt that fixed triads of deities existed in Syria--for example, at Palmyra.[222] Triads consisting of a god, goddess, and child are also known from the Hittite world.[223] But the existence of fixed triads on Phoenician soil is a debated issue.[224] Did Baᶜal-Ḥamon, Tanit, and Rešep constitute some kind of triad at Carthage?

There appears to be some evidence for this. In his theogony Sakkunyaton records that "there were born to Kronos in Peraia three sons: Kronos, who had the same name as his father, Zeus Bēlos, and Apollo."[225] This indicates that Apollo/Rešep in our treaty may well be the son of Baᶜal-Ḥamon. Since Tanit is definitely the consort of Baᶜal-Ḥamon, it may well be that the first three deities in the H/P god-list constitute a triad of husband, wife, and son.

The same grouping of Baᶜal-Ḥamon, Tanit, and Rešep may be present in *KAI* 78.3-4, which we discussed earlier. Immediately after Tanit and Baᶜal-Ḥamon comes *bᶜl mgnm*, an otherwise unknown deity. Février has tried to identify him with ʾEšmun; but his hypothesis is based on an arbitrary emendation of the name to *bᶜl *mgnn* and thus does not commend itself.[226] Now the element *bᶜl* followed by a noun in the construct often forms a divine epithet in Semitic. At Ugarit, for example, Hadad was also known as Baᶜal-Ṣapan; we have seen above that ʾEl was Baᶜal-Ḥamon at Carthage; and at Tyre Melqart was known as *bᶜl ṣr* (*KAI* 47.1), "lord of Tyre," in the second century B.C. The only feasible interpretation of *mgnm* based on attested Phoenician usage would be "shields," taking the word as the plural of *mgn*.[227] Now in Phoenicia and Ugarit Rešep bears several epithets that associate him with warfare and military equipment. Note *ršp ḥṣ* (*KAI* 32.4) and *bᶜl ḥẓ ršp* (*UT* 1001.1.3), "Rešep(, lord) of the arrow." At Ugarit he is also called *ršp ṣbʾi* (*UT* 2004.15), which most likely means "Rešep of the army."[228] W. F. Fulco notes that in Egypt Rešep is frequently portrayed carrying a shield.[229] Thus it is not unlikely that *b^c l mgnm* means "lord of the shields," that is, Rešep. *KAI* 78.3-4 would then contain the triad Tanit, Ba^cal-Ḥamon, and Rešep as we find in this first line of the H/P god-list, though in a slightly different sequence.

It remains to be seen whether this pattern of triads is continued in the god-list. But the presence of this clearly triadic group at the beginning may well indicate that the pattern of threes, so predominant throughout the list, is based on the model of a fixed divine triad--just as the pattern of twos in A/M and Sf1 is based on the model of the fixed divine pair that begins the list.

This triadic pattern, then, may be understood as an extension of the older pattern of divine pairs. In this regard H/P appears to contain something of an innovation not present in earlier prototypes such as E/B; note that the Tyrian god-list begins, like Sf1, with a divine consort pair (Bethel and ^cAnat-Bethel). This development in the treaty form is attested only in H/P, the last in the line of Near Eastern treaties, and hence is significant for the history of the treaty tradition. The god-list thus bears at one and the same time marks of great antiquity (as we shall see in detail later) as well as innovation reflecting an "updated" regrouping of the Tyrian-Carthaginian pantheon. It may have been this need to add a third member related to the supreme consort pair that necessitated the inclusion of Apollo/Rešep.

## 2. ἐναντίον δαίμονος Καρχηδονίων καὶ Ἡρακλέους καὶ Ἰολάου

*a. Daimōn Karchēdoniōn.* Of the divine names in the present section perhaps none has generated more debate than this deity, nor have so many suggestions been proposed for the identification of any of the other gods listed here. Several of these proposals are so lacking in supporting evidence that they need not be seriously considered--for example, J. M. Solá Solé's proposal of the Egyptian god Bes as the *daimōn Karchēdoniōn*[230] (hereafter *d.K.*) or Février's attempt to identify *b^c l mgnm* in *KAI* 78 with the *d.K.* H. Winckler has suggested an identification with Dido.[231] But the majority of commentators identify the *d.K.* with Gad, Tanit, or Aštart. As we shall see, it is with one of these three that the correct choice lies.

Some time ago F. Cumont[232] pointed out the connection between the Greek terms *daimōn* or *tychē* (which are related semantically)[233] and the West Semitic *gad*; all three terms denote a "tutelary genius" or "protective deity." The temple of the local *genius* or *tychē*, *tycheion* in Greek, was called *bêt gaddâ'* (lit., "the temple of the *gad*") in the Syrian world.[234] Moreover, the LXX translates Hebrew *gad* as *daimōn* in Isa 65:11. And despite the hesitations of a few,[235] *d.K.* is transparently identical with the Latin *genius Carthaginis*; it would be difficult to imagine a more straightforward Latin translation of *d.K.* or the underlying Punic phrase. Cumont also observed that although the identification of *gad* with *tychē* is attested in inscriptions, *tychē* is feminine whereas *gad* is masculine.[236] This is probably one of the reasons why a number of scholars have proposed a masculine deity, "Gad," as the *d.K.*[237]

But as a matter of fact there is no proof that *gad* was the proper name of a deity in the ancient Near East.[238] This appears to be especially clear in the Phoenician-Punic world. Prescinding from its use as a theophorous element in personal names, the word *gad* occurs twice in Punic inscriptions. In one case (*KAI* 147.2) the reading is not certain: the text seems to have *gd hšmm*; this could be understood either as "Gad of the heavens" or as "the *gad* [tutelary deity] of the heavens." But in the other documented occurrence of the name in a Punic text (*KAI* 72 B.4) the term is preceded by the article--*hgd*; this can only be translated "the *gad*." In the OT too, in the sole instance where it occurs, *gad* takes the article--*laggad*, "to the *gad*" (Isa 65:11; LXX, *tōi daimoni*). And in Palmyrene inscriptions it frequently takes the article: *gd' dy dwr'*, "the *gad* of Dura"; *gd' dy tdmwr*, "the *gad* of Palmyra."[239]

These last two expressions are quite significant, being almost identical formally to *d.K.*; in Palmyrene the city-names are *nomina recta*, whereas *Karchēdoniōn* is a gentilic, but the meaning is the same.[240] Such clearly parallel expressions should remove any doubt that in *d.K.* the first term translates Punic *gad*. Moreover, the use of the common noun in Palmyrene and the use of the article with *gd* in Punic indicate that *d.K.* is to be understood not as a proper name but as an epithet--"the *gad* of the Carthaginians." Notwithstanding Cumont's contention, the *gad* can be masculine or feminine. The *gad* of the "blessed spring" of Ephca in Palmyra, *gd' dy ᶜyn' brykt'* (*CIS* 2.3976´), was the god Yarḥibol.[241] The *gad* of Palmyra, however, *gd' dy tdmwr*, was the goddess Atargatis,[242] who is elsewhere called *Tychē Palmyrōn*.[243]

The fact that the *gad* of Palmyra is none other than the great goddess Atargatis, the chief goddess of that city, also contradicts the view that the *gad* was merely a "figure of minor account," as S. Moscati has recently styled the *d.K.*[244] In Mesopotamia too the *lamassu* (the semantic equivalent of West Semitic *gad*) was often a major god or goddess. Thus W. von Soden notes that Adad was called *lamassi māti*, "the protective deity of the land," and that Ištar was also called a *lamassu* on occasion.[245] Once the goddess Tašmetum, the consort of Nabu, is addressed as dLAMMA-*at māti*, "the protective deity of the land."[246] These designations remind us of dLAMMA uru*Ḫatti*. Although dLAMMA by itself stands for a particular deity,[247] we now know from recently published Akkadian-Hittite vocabulary lists that Akkadian *lamassu* (Sumerographic dLAMMA) was equivalent to the common noun *annari*- in Hittite.[248] Hence dLAMMA uru*Ḫatti* may mean "the protective deity of Hatti." One should note, incidentally, that in these Akkadian, Palmyrene, and (probably) Hittite constructions with a geographic designation the first element (dLAMMA, *lamassu*, *annari*-, *gaddâ*') is not a divine proper name; this again confirms the contention that the *gad* which underlies *daimōn* in H/P is also a common noun.

Having eliminated the hypothetical "Gad" as the solution to the identification of the *d.K.*, we are faced with two possibilities: Tanit or ᶜAštart. L. F. Benedetto notes that "all the sources are in agreement in making Juno the great protectress of Carthage and of Africa,"[249] although she is never directly identified with the *genius Carthaginis*. It will be remembered that Juno answers either to Tanit or ᶜAštart at Carthage. One might expect the *d.K.* to be one of these two for another reason. The epithet *d.K.* is semantically equivalent to *Tychē Karchēdoniōn*, although this latter phrase is not attested. Now the *tychē* of a city was, of course, always a *goddess*, sometimes a major goddess such as Atargatis. Since Tanit and ᶜAštart were the two most important Punic goddesses at Carthage at the time of Hannibal, we might expect one of them to be the *d.K.*

If these are the two possibilities we have to work with, it would at first glance seem an easy choice--by process of elimination. Hera almost certainly corresponds to Tanit in this god-list; and as we have seen,[250] it is rather unlikely that a deity would be mentioned a second time under a different epithet, especially in so carefully structured a list as the present one. S. Gsell too, after toying with the possibility of Tanit's being mentioned twice in H/P (as Hera and *d.K.*), rightly concludes: "the hypothesis of a double mention [of Tanit] is hardly possible."[251] Hence $^{c}$Aštart would be the *d.K.* But since there appear to be arguments in favor of identifying the *d.K.* with each of the goddesses, it will be necessary to discuss the evidence in detail.

There is no *inscriptional* evidence that directly equates either of these goddesses with the *d.K.* or, for that matter, with the *genius Carthaginis*. There is, however, a Punic text that has been understood to associate the name of Tanit with *gad*. *KAI* 72 is an inscription on a small bronze tablet discovered on the island of Ibiza. The tablet contains two inscriptions, the first dating from the fifth century B.C., the second from around 180 B.C. It is the latter text that concerns us here. Part of this text reads, *lrbbtn ltnt 'drt whgd*--"to our Lady, Tanit the mighty and the protective deity." Is *hgd* an epithet of Tanit here or does it refer to a second deity?

J. M. Solá Solé translates the phrase "and (for) the Gad,"[252] as if supplying a missing preposition (*w<l>hgd*). H. Donner and W. Röllig also understand *hgd* in this way.[253] But G. Garbini has argued that *hgd* is in apposition to Tanit, and he finds in this text an argument for seeing Tanit as the *d.K.* of our treaty.[254] He notes that before the names of individual deities in Phoenician-Punic votive inscriptions the preposition *l* is usually repeated, but that it is not repeated when the name of a deity is followed by an epithet.[255] Garbini's position has been followed by M. Guzzo Amadasi: "It is perhaps more probable that, according to G. Garbini's proposal, p. 213, we are dealing with an epithet of Tanit, parallel to *'drt*."[256]

But Garbini's argument is not conclusive. One finds the combination of an adjectival epithet (*'drt*) plus a nominal epithet (*hgd*) only here in Phoenician-Punic inscriptions. The combination sounds strange: "Tanit, the mighty and the *gad*." One would expect two adjectives or perhaps two nouns, but not this odd combination. Secondly, there are cases in which the preposition *l* is missing before a second deity in a votive inscription. For example, *KAI* 102.1 (between the third and first centuries B.C.) reads: *l'dn lb*$^{c}$*l ḥmn wrbtn tnt p*$^{c}$*n b*$^{c}$*l*--"to the Lord, to Ba$^{c}$al-Ḥamon, and (to) our Lady Tanit, 'Face-of-Ba$^{c}$al.'" One must either emend to *w<l>rbtn* or, more probably, understand the preposition *l* in *lb*$^{c}$*l ḥmn* to govern *rbtn* as well. Thus in *KAI* 72 B one may either emend to *w<l>hgd* or, more probably, understand the *l* of *ltnt* to govern *hgd* as well. One could also take *rbbtn* in the Ibiza inscription as a plural form, which would be preferable if one understands *hgd* as a separate deity. Compare *KAI* 81.1: *lrbt l*ᶜ*štrt wltnt blbnn*--"To the *Ladies*, to ᶜAštart and Tanit-in-Lebanon." Here too the *rbt* could be either singular or plural; but the presence of two divine names demands the plural reading. Hence *KAI* 72 B does not provide solid support for the thesis that Tanit was the *gad* of Carthage; in our opinion the more natural reading would understand *hgd* as a separate deity. Finally, there is no guarantee that *hgd*--whether it be connected with Tanit or not--refers to the *gad* of Carthage rather than some local protective deity.

From Dacia during the Roman period comes a Latin inscription dedicated to *Caelesti augustae et Aesculapio et genio Carthaginis et genio Daciarum* (*CIL* 3.993). According to R. Dussaud the *genius Carthaginis* here is not equivalent to the *d.K.*[257] But Solá Solé is undoubtedly correct when he objects, "It is difficult to believe . . . that this *genio Carthaginis* has nothing to do with the *daimōn* of Hannibal's oath, as R. Dussaud maintains."[258] In any case if the *genius Carthaginis* is in fact the *d.K.*, what does the inscription tell us about the relationship between the *daimōn* and Tanit or ᶜAštart? Again the interpretation is not certain. The chief question is, who is *Caelestis augusta*? Solá Solé and others believe that this refers to ᶜAštart.[259] It is true that ᶜAštart bore the epithet *caelestis/ourania*.[260] But it is also true that "in the first centuries of our era, the usual name of Tanit Pene Baal was not *Ops* or *Nutrix* but *Caelestis*."[261] This was Tanit's standard epithet in Spain.[262] At the sanctuary of Baᶜal-Ḥamōn and Tanit in Thinissut there is a statue of a goddess standing on a lion;[263] in all likelihood the goddess represented here is Tanit.[264] On the back of the statue there is a Latin inscription which begins with the letters *C A S*; judging from other Latin inscriptions at the same sanctuary, such as *SATVRNO AVG SAC*, which clearly stands for *Saturno aug(usto) sac(rum)*--"the sanctuary (dedicated) to august Saturn [i.e., Baᶜal-Ḥamon]"[265]--*C A S* should stand for *C(aelesti) a(ugustae) s(acrum)*[266]--"the sanctuary (dedicated) to Caelestis augusta." Here we seem to have some direct evidence for the identification of *Caelestis augusta* with Tanit. Thus this inscription cannot be used to prove that ᶜAštart was distinct from the *genius Carthaginis* (and hence from the *d.K.*); if anything, it is more likely that *Tanit* is *Caelestis augusta* here and therefore not the *genius*.

We now turn to evidence from *coins*. A coin of special interest to us is one struck by Q. Caecilius Metellus Pius Scipio, the head of the

Pompeian party in Africa, in the middle of the first century B.C. Depicted on the coin is a *Mischwesen* with the head of a lion and the body of a woman; two large wings are folded in the lower front of the image.[267] The figure is accompanied by three letters, *G T A*, which are usually taken to stand for *Genius Terrae Africae*. Since Carthage was the most important city in Africa at this time, it is not unreasonable to assume that this figure is also the *genius Carthaginis* or *d.K.*, although this has yet to be established beyond question.

Who is the lion-headed goddess here? The lion was associated with a number of deities in the Canaanite-Phoenician world, including ᶜAštart and Tanit. In Egypt the three great Canaanite goddesses ʾAšerah, ᶜAnat, and ᶜAštart are portrayed standing on a lion; this iconography is not unrelated to the fact that the three tended to fuse in Egypt.[268] But of the three it is ᶜAštart who has the strongest and oldest association with the lion figure. We noted earlier that Cross refers to "the old epithet of Asherah *labiʾt(u)*."[269] But there is no evidence that ʾAšerah ever bore this epithet either at Ugarit or elsewhere in Syria-Palestine. Interestingly Cross footnotes this association of ʾAšerah and *labiʾt(u)* with a reference to an earlier article of his; but in that article he identifies ᶜ*Anat*, not ʾAšerah, as the "Lion Lady."[270] And although he is no doubt correct in associating the title with a war-goddess (here ᶜAnat), again there is no evidence that the deity in question is ᶜAnat--or ʾAšerah, for that matter. The only evidence of a "Lion Lady" in Palestine comes from proper names. The name ᶜ*bdlbʾt*, "Servant of the Lion Lady," is found in a twelfth/eleventh-century B.C. inscription (*KAI* 21) and in a text from Ugarit (*CTA* 119[321].3.38).

But it is difficult to see how Cross concludes that this goddess must be ᶜAnat or ʾAšerah without adducing evidence in favor of these identifications. There are, in fact, stronger grounds for suggesting that the Lion Lady is actually ᶜAštart. Donner and Röllig[271] identify the goddess *lbʾt* with ᶜAštart or ᶜAnat. F. (Quest-)Gröndahl[272] takes the name *lbʾit* at Ugarit to refer "probably" to ᶜAštart. One should note here the Akkadian epithet *labbatu/lābatu*, cognate with West Semitic *labiʾt(u)*. It is significant that this Akkadian word is attested only as an epithet of Ištar,[273] the Mesopotamian equivalent of ᶜAštart. The Akkadian proper name ᵈ*Ši-la-ba-at*, "She [i.e., Ištar] is a lioness,"[274] turns up, interestingly, in Akkadian documents from Ugarit.[275] Thus of the three chief Canaanite goddesses, it is most likely ᶜAštart who is the "Lion Lady."

But there is more direct evidence of a connection between the lion-headed figure on the Scipio coin and ᶜAštart. In the sanctuary excavated at Bir Bou-Rekba (Thinissut) were discovered several statues of a lion-headed goddess, almost identical in appearance to the figure on the coin.[276] Now there is no doubt that the temple is that of Baᶜal-Ḥamon and his consort Tanit, as is explicitly stated on a marble plaque (*KAI* 137) discovered at the site. Thus A. Merlin identifies the figure with Tanit.[277] Cross seems to imply the same thing when he speaks of the lion-headed statue as "a different tradition of the *Qudšu* [i.e., ʾAšerah] iconography. . . .";[278] for Cross ʾAšerah is Tanit at Carthage. But the presence of these statues in a sanctuary of Tanit should not be taken as proof that the figure is an image of this goddess. Merlin himself admits the presence of statues of goddesses other than Tanit in the sanctuary. He speaks, for example, of "the series of Baalits who form Baal's cortege in our sanctuary and who are all grouped around Tanit. . . ."[279] One of these is a goddess wearing armor, whom Merlin identifies as Athena. Thus Tanit is not the only goddess represented here. Why, then, should one insist that the lion-headed goddess is Tanit? There is no direct evidence for such an association. We have just seen that a much stronger case can be made, at least epigraphically, for the leonine character of ᶜAštart. In addition there is iconographic evidence. Merlin[280] and Cross[281] acknowledge that the lion-headed figure is an adaptation of the Egyptian war-goddess Sekhmet, who is frequently depicted in this way.[282] Now there is a bas-relief from the temple of Horus at Edfu depicting a lion-headed goddess; the accompanying inscription explicitly identifies her as ᶜAštart.[283] R. du Mesnil du Buisson believes that in the well-known cylinder seal from Bethel[284] ᶜAštart is portrayed with the head of a lioness, although this is not clear.[285]

From all of this it is evident that the lion-headed goddess at Thinissut is almost certainly ᶜAštart, not Tanit. Here ᶜAštart merely shares the sanctuary of Tanit, a phenomenon attested in the dedicatory plaque from a temple of ᶜAštart and Tanit at Carthage (*KAI* 81). Since the figure in the Thinissut sanctuary is identical with the image on the coin of Scipio, the *G T A* is likewise ᶜAštart rather than Tanit. And if the meaning of these letters has been correctly understood, ᶜAštart is the *Genius Terrae Africae* and therefore (perhaps) the *d.K.* as well.

It is also significant that on coins from Phoenicia and Palestine ᶜAštart seems to be represented as the *tychē* of the city, the "city-goddess," with great frequency, so that it is not easy to tell one goddess from the other. "This difficulty of distinguishing between Tyche and Astarte confronts us in nearly all the cities of the Phoenician coast," claims G. F. Hill.[286] And yet "there can be no doubt that the Tyche-like goddess whom we see endowed in all the maritime cities with maritime attributes, such as the prow of a vessel, a naval standard, or an aphlaston, is Astarte. . . ."[287] On some coins from Palestine this figure is accompanied by a dove,[288] which is associated elsewhere with ᶜAštart.[289] This same figure appears on an intaglio found at Sidi-Daoud in Tunisia.[290] If ᶜAštart is so widely represented as the city-goddess (*tychē poleōs*) on coins from Phoenicia, this would seem to point to her being the *tychē* of Carthage--that is, the *d.K.*

At this point it may be helpful to mention a few facts about the image of the "city-goddess." This figure appears with great frequency on coins and reliefs in the Greco-Roman Near East. In most cases the city *tychē* was patterned on the famous bronze statue of the city-goddess of Antioch-on-the-Orontes fashioned by Eutychides of Sicyon at the behest of the city's founder, Seleucus Nicator. The statue dates from shortly after 300 B.C.[291] and a number of copies have survived.[292] The goddess is depicted as seated on a rock, with one foot upon a swimming figure, the river-god (representing the Orontes); in her right hand she holds a sheaf of grain. She wears a seemingly unusual piece of headgear, an object variously called a "city-crown," "turret-crown," or "walled-crown"; it is recognized by a series of "battlements" or turrets at intervals around the crown, so that the object resembles a miniature replica of a walled city. This city-crown is generally taken to be one of the most important identifying features of the city-goddess.[293] The aspect of the goddess with one foot on the river-god is also a distinctive trait of the *tychē*, at least in cities located on rivers;[294] in the case of coastal cities this feature is modified so that the city-goddess is shown with one foot on the prow or rudder of a galley.[295] Sometimes the figure merely holds a cornucopia and rudder.[296] Thus we may single out three features that are widely held to be distinctive of the city-goddess: (1) the city-crown; (2) the foot on a river-god or ship; (3) the cornucopia and rudder.

Is Tanit or ᶜAštart ever depicted with one of these features of the city-goddess? We have seen that ᶜAštart is thought to be widely identified with the *tychē*. But one must admit that this identification is not entirely certain. It has been noted that the attributes of the *tychē* and other goddesses occasionally fuse so as to designate a single all-embracing goddess, the so-called *Tychē Pantheia.*[297] In such cases it would be inaccurate to speak of the city-goddess as ᶜAštart, since this identification would too narrowly limit the personality of the pantheistic *tychē*. So it is necessary to look for further data in order to resolve the problem.

There is an artifact from North Africa which some have taken as conclusive evidence that Tanit was regarded as the city-goddess of Carthage--a silver headband discovered at Batna in Algeria,[298] which probably served as a priestly diadem.[299] The two central figures on the band are those of a god and goddess. The god has ram horns and is therefore certainly Baᶜal-Ḥamon, under the aspect of Zeus/Jupiter Ammon. The similarity between the names *ḥmn* (pronounced Amūn)[300] and Amōn/Amūn led to the identification of the two gods,[301] so that Baᶜal-Ḥamon came to assume the characteristic ram horns[302] of the Egyptian god. The identification of the goddess is also certain: she is Tanit. This is apparent not only from the fact that she is depicted beside Baᶜal-Ḥamon (and thus as his consort) but also from the symbols on the headband itself. These include almost every symbol thought to be associated with Tanit: the "sign of Tanit," the dove, the dolphin, the caduceus, etc. Now what is most significant about this image of Tanit is that *she is wearing the city-crown.*[303] From this fact Gsell[304] and Baudissin[305] conclude that Tanit is the city-goddess. This would seem to indicate that Tanit, not ᶜAštart, is the *d.K.*

But data recently brought to light render this argument inconclusive. Among the personal effects of Queen Aḫatmilku of Ugarit itemized in an inventory list (RS 16.146 + 161) we find the following entry: 1 URU GUŠKIN KI.LÁ.BI *2 me-at 15*--"one 'city of gold': weight, 215 (shekels)."[306] J. Nougayrol provisionally understood this peculiar term to denote a "city (-crown)."[307] This interpretation has now been fully substantiated. S. M. Paul was the first to note a connection between this URU.GUŠKIN ("city of gold") and a term encountered in the rabbinic writings, *ᶜyr šl zhb*--likewise meaning "city of gold."[308] From the rabbinic references it is quite clear that the object in question is in fact a crown of gold; it is at times specifically designated as such (*ktr šl zhb*).[309] This crown was "an expensive piece of jewelery worn only by women of high status."[310] Paul also notes that among the Hittite reliefs at Yazilikaya certain goddesses appear to be wearing a "battlemented/turreted crown";[311] this observation has been confirmed by the Hittitologist H. A. Hoffner.[312] Hoffner also cites a Hittite text mentioning 1 URU$^{LUM}$ KÙ.BABBAR--"one 'city of silver'," to be made for certain Hittite deities.[313] It is evident from Queen

Aḫatmilku's inventory as well as the rabbinic texts that the city-crown was a highly prized--and expensive--item of fashionable apparel in the ancient Near East for several millennia. Paul observes that crowns of this type appear in representations of queens from Elam and Assyria;[314] in his most recent article on this subject he has included an illustration of a panel from the famous third-century A.D. synagogue at Dura-Europos, which shows Queen Esther wearing a golden city-crown.[315] Apparently the crown was originally an item of royal or noble apparel; but it is only to be expected that deities, lest they seem to be less stylish than mortals, would likewise be portrayed wearing these crowns.

The "city-crown," then, was worn not only by various gods and goddesses but also by royalty for a very long time both before and after Eutychides produced the famous statue of the Tyche of Antioch. This means that, contrary to the prevailing opinion, the city-crown is *not* a distinctive feature of the city-goddess. It appears on the Tyche of Antioch not because *city*-crown denotes *city*-goddess, but simply because fashionable queens and goddesses of the Near East often wore it. In the absence of other specifying features of the *tychē*, then--the foot on the river-god or prow, the cornucopia--it is no longer admissible to conclude that any Syro-Palestinian figure wearing the city-crown must be a city-goddess. Thus the fact that Tanit is shown wearing this object on the Batna headband cannot be taken as evidence that she was regarded as a city-goddess. And if this is so, the headband can no longer serve to support the claim that Tanit was the *d.K.*

Although we must still await unambiguous proof that ᶜAštart was in fact the *d.K.*, the weight of evidence presented here makes this identification probable. The *d.K.* can hardly be Tanit, since the latter appears in the first line under the name Hera; on the other hand it would be very unlikely indeed that so important a deity as ᶜAštart be omitted from an official list of the deities of Carthage. A final and important piece of evidence is the fact that ᶜAštart is named in the Tyrian god-list in E/B (*ᵈAs-tar-tú*). In fact we shall see directly that in H/P she is associated with the same gods with whom she is listed in E/B. However, in the latter treaty she is the last named of the high gods, whereas she is named first in the second line of the H/P list, under the epithet *d.K.* As we suggested earlier, the position of *d.K.* and therefore of ᶜAštart as the first entry of the second line of the god-list may possibly reflect something of the structure of god-lists from the Hittite treaty tradition, where "the protective deity (of Hatti)" begins the second section.

*b. Hēraklēs*. If the *d.K.* is one of the more controversial deities listed here, Heracles is surely the most readily identifiable. This god is certainly Melqart, who is called "the lord of Tyre" (*bᶜl ṣr*) in a second-century B.C. bilingual inscription (*KAI* 47). The equivalence of the two gods is known not only from the inscriptions but from the testimony of Sakkunyaton as well.[316] It is known that Melqart had a temple in Carthage,[317] and that the Carthaginians had the practice of sending a tenth of their income to his temple in Tyre annually.[318] Moreover, the importance of this god at Carthage is indicated by the fact that Melqart (*mlqrt*) is by far the most common theophorous element in Punic proper names.[319] Hence virtually all commentators on this treaty agree in identifying Heracles here with Melqart.[320] As regards the Tyrian treaty tradition, note that Melqart is invoked as a divine witness in E/B (ᵈ*Mi-il-qar-tu*).

*c. Iolaos*. The identity of this deity is more problematic. Unlike the other divine names we have encountered so far in this section, this god (or, more accurately from the Greek viewpoint, divinized hero) has no Punic counterpart with whom he is regularly equated.[321] This is the chief source of difficulty in ascertaining the identity of Iolaos. In Greek mythology he is "the nephew, faithful friend, and companion of Heracles,"[322] sometimes acting as his charioteer and armor-bearer. Heracles is said to have established a sanctuary for him in Sicily,[323] and he was worshipped beside Heracles in Thebes.[324]

It is precisely this association with Heracles that explains the choice of the name Iolaos here in the god-list. Just as the translators of the treaty chose the name Zeus for Baᶜal-Ḥamon (instead of the standard equivalent, Kronos) in order to show that the first two deities are related as the supreme consort pair, so too here the name Iolaos was undoubtedly selected to indicate that the Punic deity in mind is intimately associated with Melqart, as was Iolaos with Heracles. This does not necessarily mean that the Punic deity had the same relation to Melqart that Iolaos had to Heracles. Rather, there was no deity more popularly associated with Heracles than Iolaos, and so the choice would almost inevitably fall to him.

The first question to be asked then is, which Punic god was closely associated with Melqart? Again we come to two possibilities: Ṣid and Ešmun.

The name of the god Ṣid (*ṣd* in Punic) is connected with Tanit and also Melqart: *ṣd-tnt* (*CIS* 1.247-49) and *ṣd-mlqrt* (*CIS* 1.156). The precise relationship of two divine names in such constructions (which appear to be construct chains) is ambiguous. Sometimes the relationship is conjugal; this is unquestionably the case in a name such as ᶜAnat-Bethel in E/B or ᶜAnat-Yahu in the Aramaic papyri, and most probably too in the case of Milkᶜaštart.[325] Or, the relationship could be syncretistic. This might be true of *ᶜnt.wᶜṯtrt* at Ugarit (later Atargatis)[326] and perhaps also of *tntᶜštrt*. Alternatively, a combination like *tntᶜštrt* might mean nothing more than "Tanit (worshipped alongside) ᶜAštart." *KAI* 81 is evidence of the veneration of Tanit and ᶜAštart in the same temple complex. And although we have indicated that *mlkᶜštrt* probably denotes a conjugal relationship, it is not impossible that here too the phrase might (also) mean "Milk (worshipped alongside) ᶜAštart"; for *KAI* 19.4 speaks of a dedication *lᶜštrt bʾšrt ʾl ḥmn*--"to ᶜAštart in the sanctuary of the god of Ḥamōn [ = Milkᶜaštart]."[327] Similarly Baudissin thinks that *ṣd-tnt* specifies Ṣid as worshipped in a temple belonging to Tanit.[328] This specification of one deity by another is not attested in the earlier period; but functionally it is not unrelated to earlier designations in which the deity is specified by a cult-place. Thus Ištar of Nineveh means "Ištar, namely, Ištar as she is worshipped in her temple at Nineveh," distinct from Ištar of Arbela, etc.; likewise *ṣd-tnt* might mean "Ṣid, namely, Ṣid as he is worshipped in the temple of Tanit."[329]

R. Dussaud has argued that certain constructions of this type denote a filial relationship; for example, *ʾšmn-mlqrt* would mean "ʾEšmun(, son of Melqart,"[330] and *mlqrt-rṣp* would mean "Melqart(, son of) Rešep."[331] F. Vattioni has rightly rejected this explanation of the syntax of the construct chain here.[332] For one thing, the order of the deities in such constructions often appears to be optional: *mlqrt-ṣd* (*CIS* 1.256), but *ṣd-mlqrt* (*CIS* 1.156); *ʾšmn-mlqrt* (*CIS* 1.245), but *mlqrt-ʾšmn* (*CIS* 1.16); and *ʾršp-mlqrt* (*KAI* 72 A) in addition to *mlqrt-rṣp* (*CIS* 1.150). This is not to say that in the case of certain of these double divine names a father-son (or mother-son) relationship may not exist; but such a relationship is in no way indicated by the syntax. In other words, Dussaud's insight may be correct (in some instances); but he bases his understanding of the names on an incorrect interpretation of the grammar involved.

Hence the divine name *ṣd-mlqrt* expresses some close association between the gods Ṣid and Melqart. The association may have its origins in a mythological event, in a likeness of nature and/or function, or in a coveneration in the cult--it may even be filial. Some evidence for the last interpretation may be found on an engraved razor from Carthage.[333] One side depicts Heracles/Melqart (identified by the lionpelt worn on the head) and the other, a god wearing a calathos-like crown of feathers, piercing a conquered foe with a lance. Now a coin of Atius Balbus (59 B.C.) likewise shows a god with a crown of plumes, holding what may be a lance.[334] The inscription on the coin reads *SARD PATER*, that is, *Sardus Pater*. The second-century B.C. Greek writer Pausanias[335] relates that the Libyans had occupied Sardinia under the command of their leader *Sardos*, son of *Makēris*.[336] He adds that the latter was surnamed Heracles by the Egyptians and Libyans. Thus in all likelihood *Makēris* is a deformation of the name Melqart.[337] This tradition, if accurate, preserves the reminiscence of a father-son relationship between Melqart and Sardus. There is now evidence that in the Punic world Ṣid was identified with *Sardus Pater*. The Italian excavation of Antas in Sardinia has unearthed a temple that reveals three phases of construction: (1) an archaic Punic period (sixth to fifth centuries B.C.), (2) a late Punic period (third century B.C.), and (3) a Roman period (second to third centuries A.D.).[338] The inscriptions from the Punic periods are dedicated to the god Ṣid, while the Latin dedication names Sardus Pater.[339]

> The Punic inscriptions from Antas show that at this site, during a period that can be dated approximately between the fifth and the second century B.C., the antecedent of the Roman *Sardus Pater* was a Phoenician god. The information from classical sources on the relationship between *Sardus Pater* and Heracles might thus possibly be related to the connection established between *ṣd* and *mlqrt* in the Carthaginian dedicatory inscription *CIS* I, 256.[340]

But the god ʼEšmun (*ʼšmn* in Punic) is also associated with Melqart. In the Punic world one finds *ʼšmn-mlqrt* (*CIS* 1.16) and *mlqrt-ʼšmn* (*CIS* 1.16,23-28); compare *ṣd-mlqrt* and *mlqrt-ṣd*. Moreover, statements by Greek authors may also suggest a connection between the two gods. Athenaeus[341] cites the following excerpt from the writings of Eudoxus of Cnidus (fourth century B.C.):

> He [Eudoxus] says that the Phoenicians sacrifice quails to Heracles because when Heracles, son of Asteria and Zeus, went to Libya and was killed by Typhon, he came back to life when he perceived the aroma of the quail Iolaos had brought him.

The same legend is reported by Zenobius:[342]

> Euxodus says that the Tyrian Heracles was killed by Typhon; but that Iolaos, doing everything [he could] to raise Heracles, roasted a live quail (which Heracles liked); Heracles came back to life from [smelling] the aroma.

Both of these accounts make it clear that this legend really concerns Melqart rather than the Greek hero: according to Athenaeus' version the story comes from the Phoenicians, and according to Zenobius the tale concerns the Tyrian Heracles. Hence the Iolaos mentioned here must likewise be a Phoenician Iolaos, that is, some Phoenician hero or deity associated with Melqart. The fact that Iolaos is cast in the role of a *healing* god or hero--which is not true of the Greek Iolaos--reminds one of Asklepios,[343] the god of healing par excellence, with whom 'Ešmun is more commonly identified.[344] No other Phoenician god is so explicitly connected with healing except perhaps Šadrapa.

Since the Punic god associated with Heracles here is specifically named *Iolaos* (rather than *Sardus*),[345] it would appear that it is this "healing" Iolaos that we find in H/P. Baudissin's arguments[346] for the identification of the Iolaos in our treaty with 'Ešmun are generally convincing. One piece of evidence he mentions[347] tips the scales decisively in favor of 'Ešmun, although he simply lists it as one argument among many; this is the fact that Melqart and 'Ešmun (in that sequence) are named in E/B. In fact they are named *together* and share the curses invoked after their names; this is the only pair of Phoenician deities named in the god-list except, of course, the supreme consort pair. What is important here is that this explicit association of the two deities occurs in the Tyrian god-list in E/B. Furthermore, in E/B Melqart and

ʾEšmun ($^{d}$*E!-su-mu-nu*) are followed by ʿAštart, just as ʿAštart (as the *d.K.*) immediately precedes Heracles/Melqart and Iolaos/ʾEšmun in this line of H/P. Given the general conservatism of this tradition and bearing in mind the arguments for identifying the Punic "Iolaos" with ʾEšmun, it is all but certain that the Iolaos of our treaty is in fact the god ʾEšmun. He is not called Asklepios here because the treaty tradition is not directly interested in his role as healer but rather in his association with Melqart; this association is evoked by the name Iolaos, not Asklepios.

Our conclusion is that the second line of deities in this god-list, like the first, does in fact consist of a *triad* of deities: a consort pair plus a third god related to one or both of this pair. Although Athenaeus[348] and Cicero[349] agree in making Melqart the son of "Asteria" ( = ʿAštart?)[350] the evidence is rather impressive for the consort relationship of the two deities.[351] It is hard to deny, first of all, that Milkʿaštart is none other than Melqart.[352] The former name appears on a dedication from Cadiz (*KAI* 71), where Melqart, *Hercules Gaditanus*, was the major deity. We have already seen that ʿAštart shared a sanctuary with Milkʿaštart at Umm el-ʿAwāmīd (*KAI* 19.4). The two are named together in inscriptions from Leptis Magna (where ʿAštart is called *Leukothea*)[353] and even in inscriptions from Roman Britain.[354] Just how ʾEšmun was related to Melqart--beyond the fact that he "revived" Melqart--is not entirely clear. One should note here the worship of Asteria, Heracles, and Asklepios at ʿAmman;[355] the sequence of deities named in this triad is precisely the same as in this line of the H/P god-list.

It is now rather obvious that this part of the god-list in H/P represents an expansion of the older dyadic pattern, as seen in A/M, Sf1, and parts of E/B, into a triadic pattern. The consort pair (taken over from the older model) forms the nucleus of this arrangement, although as we shall see only these first two lines contain consort pairs. Yet the triadic arrangement continues into the second major sub-division of the god-list.

### 3. ἐναντίον Ἄρεως, Τρίτωνος, Ποσειδῶνος

Less has been written on the identification of these three deities than any of the others in the god-list. Benedetto devotes merely one line to them.[356] In fact only one article has been produced on the identification of this group of gods.[357] F. W. Walbank adequately summarizes the state of the question: "On present evidence the identity of Poseidon, Triton, and Ares remains uncertain.[358]

As in the case of Iolaos, there are no Punic deities with whom these gods are regularly identified; but unlike the case of Iolaos, we have little or no material from ancient authors to help us here. It would be useful to begin, then, with a brief description of the nature and function(s) of the Greek deities.

Ares was the Greek god of war[359] par excellence, identified at Rome with Mars. According to H. J. Rose he was, in fact, the deification of the warlike spirit. Son of Zeus and Hera, he was never a particularly popular god and had a "ferocious and unlovely" temperament.[360] His symbols were the spear and the burning torch.[361]

Triton was "the merman of Greek, or rather pre-Greek mythology."[362] He was the son of Poseidon and Amphitrite and lived with them in a golden palace in the depths of the sea.[363] "His special attribute is a twisted sea-shell, on which he blows, now violently, now gently, to raise or calm the billows."[364] Gradually the god's individuality seems to have diminished, so that one finds "Tritons," who generally appear "as a decoration of sea-pieces or other works of art. . . ."[365]

Poseidon, son of Kronos and Rhea, was the god of the sea and flowing waters in general.[366] "Being the ruler of the sea (the Mediterranean), he is described as gathering clouds and calling forth storms, but at the same time he had it in his power to grant a successful voyage and save those who are in danger. . . ."[367] Since the sea was believed to surround the earth, Poseidon was called *gaiēochos*, "he who holds the earth"; he could even cause earthquakes, and hence was known as *kinētēr gēs*, "mover of the earth." He was regarded as the creator of the horse.[368]

Since the last two gods are connected with the sea, it is probable that one or more of the underlying Punic gods also has marine associations. All three deities are masculine, which implies that the same is true of the corresponding Carthaginian gods. The pattern established in the first two lines, namely, that of a consort pair plus a third, related god, is not present in this line, although the triadic structure is nevertheless continued.

In the discussion of the foregoing two lines of gods we were led to posit a direct connection between the list of Tyrian oath-gods in E/B and the divine witnesses in our treaty; thus far all of the latter (except Apollo/Rešep) are to be found in the E/B god-list. In this earlier list there are three remaining gods, listed together and sharing the curses that follow their names: Baᶜal-Šamem, Baᶜal-Malage, and Baᶜal-Ṣaphon.

As we shall see, at least one of these--in addition to his function as a storm god--has marine associations. The question therefore arises whether there could be some connection between these three remaining deities in the E/B god-list and the last three gods in this section of H/P. Once again the extreme conservatism of the treaty tradition would seem to point in this direction.

In discussing this line we shall depart from the format followed up to now. Rather than treat these gods in the order in which they occur, we shall proceed from what we consider the more easily identifiable deities to those more difficult to identify. The sequence will thus be Poseidon, Ares, Triton.

*a. Poseidōn.* R. Dussaud, in the only extensive work on the identification of these gods, equates Poseidon with Yamm ("Sea"). He believes that the merman-god depicted on certain Phoenician coins is Yamm,[369] and that Yamm here is equivalent to Greek Pontos or Poseidon.[370] One could object that the merman-god on these coins could just as easily represent a Punic version of Triton. On the other hand, it seems that Yamm did have a temple at Carthage;[371] and Yamm certainly corresponds closely to Poseidon in his role as the sea-god par excellence.

But one of the problems involved in accepting the equivalence of Poseidon and Yamm here concerns the rank of Yamm in comparison to that of other high Canaanite-Phoenician gods. Since the "high gods" section of the H/P god-list--the first three lines--contains only nine names, one would not expect a god of minor rank to be included at the expense of higher gods, especially if some of the latter are attested elsewhere in the treaty tradition. There is no inscriptional evidence from Phoenicia that would lead us to believe that Yamm was a major deity in the first millennium B.C.; hence it is not surprising that he is not mentioned in the E/B list of Tyrian gods. We have seen that "the great sea" (A.AB.BA. GAL, Hittite *šalliš arunaš*) occurs regularly in the Hittite treaties, but always in the last section, where the divinized natural elements are listed--never among the high gods.[372] It is true that in the "pantheon list(s)" from Ugarit *ym* occurs;[373] but he is mentioned just after the *pḫr ʾilm*, "the assembly of the gods." This phrase or its equivalent is found in short god-lists within Phoenician inscriptions, where it appears to form a summary phrase that concludes a list of high gods (*KAI* 4.4: *wmpḥrt. ʾl.gbl*; 26 A III.19: *wkl dr bn ʾlm*; 27.12: *wrb.dr kl.qdšm*).[374] Moreover, in the Ugaritic pantheon list(s) *ym* is followed by the divinized incense burner (Ugaritic *ʾuṯḫt*, Akkadian $^{d.dug}$BUR.ZI.NÍG.GA) and the divinized lyre (Ugaritic *knr*, Akkadian $^{d.gis}$*ki-na-rum*), which could hardly be considered major oath-deities. All of this points to the unlikelihood--though not impossibility--of Yamm's being listed in this section of H/P as a major oath-god.

According to another point of view Poseidon here could be equivalent to Ba$^c$al(-Ṣaphon). This was argued over a century ago by F. C. Movers: "there can be no question that the sea-god whom the Carthaginians venerated . . . was the god of the Phoenician sea-farers or Baal in his character as sea-god."[375] Now here Movers is speaking of the "Libyan-Phoenician Poseidon,"[376] so that he appears to be identifying Ba$^c$al with Poseidon. This view has been cited favorably by Benedetto.[377] Albright associates the two when he says, "Hadad was himself in a general way the storm-god, but Baal-ṣaphôn was the marine storm-god *par excellence, like Greek Poseidon.*"[378] Ba$^c$al-Ṣaphon was later known as *Zeus Kasios.*[379] Now Zeus Kasios "was particularly connected with sea-faring and mariners . . ."[380] and was at times "worshipped in the form of a bark."[381] But how did Ba$^c$al-Ṣaphon, identical with Hadad at Ugarit, take on marine characteristics? This came about because of the geographical position of Mount Ṣaphon ( = Jebel el-Aqra$^c$):[382] "Since this mountain, because of its location on the shore, was a promontory which served mariners as a beacon, the 'Lord of the North' [i.e., Ba$^c$al-Ṣaphon] became a patron deity of mariners."[383] Albright also sees a connection between sea-faring and Mount Kasios ( = Mount Ṣaphon) in the writings of Sakkunyaton; the latter claims that the Dioscuri or Cabiri invented the first boat[384] and that "their descendants . . . made barges and boats, and when they were cast up (after shipwreck) on *Mount Casius* they built a temple there."[385] This role of Zeus Kasios or Ba$^c$al-Ṣaphon as protector of sea-voyagers also parallels what we have seen of Poseidon, who could grant safe voyages to sea-farers; in addition, both deities were associated with storms at sea. It will be remembered that in E/B Ba$^c$al-Ṣaphon clearly functions as a *marine* storm-god.[386]

Concerning Ba$^c$al-Ṣaphon Albright further notes, "In his honour temples were built and ports were named along the Mediterranean litoral as far as Egypt. . . ."[387] In this connection it is interesting to recall the surviving Greek text recounting a voyage of Hanno of Carthage, in which the author narrates: "Sailing thence we came to Soloeis, a Libyan *promontory* covered with trees. There we founded a temple to *Poseidon*."[388] Given the fact that the temple was built in connection with a major sea-going expedition[389] and on a promontory (cf. Ṣaphon), it would not be improbable that the Punic deity to whom the temple was dedicated was Baᶜal-Ṣaphon, not Yamm. If so, we would have in this account indirect confirmation of the equivalence of Baᶜal-Ṣaphon and Poseidon.

Finally, returning once more to the god-list from E/B, it is significant that Baᶜal-Ṣaphon is the last of the group of three gods mentioned there (ᵈ*Ba-al-ṣa-pu-nu*), just as Poseidon is the last god named in the present triad. When one considers the connection of the deities treated so far with those in E/B, this may not be accidental. Thus we conclude that in our text Poseidon is Baᶜal-Ṣaphon.

<u>*b. Arēs*.</u> Dussaud has identified the Ares of our treaty with Baᶜal (-Hadad). He appears to base this equivalence simply on the fact that "at Liban Ares (Baᶜal) is the enemy of Adonis (Môt). . . ."[390] But there is abundant evidence that in fact Baᶜal-Hadad, together with his sister ᶜAnat, was one of the "warrior deities par excellence in the Ugaritic texts."[391] Since this is so obvious from a reading of the Ugaritic literature concerning Baᶜal, we shall mention here only a few of the factors that clearly characterize him as a war-god. One of Baᶜal's standard titles at Ugarit is *ʾalʾiyn*, which Albright has shown to be related to the phrase *ʾalʾiy qrdm* (*CTA* 3[ᶜNT].3.11, etc.)--"I prevail over the heroes/warriors."[392] This is obviously the epithet of a war-god. The same is true, according to P. Miller, of the epithet *rkb ᶜrpt*, "the rider/driver of the clouds"; for here "the clouds are the war chariot of the storm god as he goes to do battle."[393] The epic literature from Ugarit portrays Baᶜal as one who vanquishes the sea-god (Yamm) and battles dreaded Death (Môt). The iconography of Baᶜal at Ugarit also portrays him as a storm-warrior, with a lightning bolt in one hand and a war mace in the other.[394]

Certainly Baᶜal qualifies, then, as a war-god par excellence and would answer in this respect to the Greek Ares. The only other likely contender among the Canaanite-Phoenician gods for this title is Rešep, with epithets like *ršp ṣbʾi* ("Rešep of the army" or "Rešep the soldier"), *bᶜl ḥẓ ršp* ("Rešep, lord of the arrow"). Rešep was also well-known in Egypt as a war-god and patron of the king beginning in the eighteenth dynasty.[395] But as we remarked earlier, it is quite unlikely that Rešep, who is almost certainly Apollo in this god-list, would be listed a second time.

It appears that Dussaud's proposal that in our text Ares is Hadad has much to commend it. Now we have argued that Hadad is Baᶜal-Šamem,[396] who occupies the first place among the triad of gods in E/B, as does Ares in this line of H/P. To be sure, Baᶜal-Šamem is most frequently identified with the storm-god Zeus among Greek deities.[397] But in the H/P god-list Zeus has already been identified (in his function as head of the pantheon) with Baᶜal-Ḥamon. Therefore another Greek divine name had to be chosen for Hadad/Baᶜal-Šamem, and Ares is an appropriate choice. Besides, it would be most unusual for so important a Carthaginian deity as Baᶜal-Šamem to be entirely omitted from this official list of the highest state gods.

But can Ares be Hadad if Poseidon is Baᶜal-Ṣaphon? It is clear from the Ugaritic texts that Hadad and Baᶜal-Ṣaphon were one and the same deity. Since we do not expect the same god to be named twice in the list--and certainly not in the same line!--this fact would at first seem to mean that the first deity (Ares in H/P, Baᶜal-Šamem in E/B) could not be Hadad. But in Phoenicia it is rather clear that Hadad/Baᶜal-Ṣaphon has undergone a development since the time of the Ugaritic texts. Albright seems to imply some differentiation between the two gods when he says, "Hadad was himself in a general way the storm-god, *but* Baal-ṣaphôn was the *marine* storm-god par excellence. . . ."[398] The evolution we see from Hadad to (Phoenician) Baᶜal-Ṣaphon to Zeus Kasios is that from a storm-god to a marine storm-god to a (predominantly) marine god. The character of Baᶜal-Ṣaphon in Phoenicia had changed significantly in relation to Hadad, so that the two could no longer be considered the same deity. The phenomenon of a god's epithet achieving an existence and a cult independent of the original deity is not unknown in the ancient Near East.[399] The same evolution did not take place, however, in the case of Baᶜal-Šamem, also apparently an epithet of Hadad; for in this instance the epithet placed emphasis on the *celestial* nature of the deity, thus giving prominence to his function as a storm-god. Hence, unlike Baᶜal-Ṣaphon, Baᶜal-Šamem never became distinct from Hadad in Phoenicia. If this line of reasoning is sound, we may conclude that Ares in our treaty is in fact Hadad, who is Baᶜal-Šamem.[400] As we noted earlier, the fact that this third line or section of the high gods begins with the name of a god who is explicitly a war-deity may reflect the arrangement in the Hittite treaty tradition, where the third section of high gods likewise consists of war-gods.

*c. Tritōn.* If, as we have argued, the first and last gods in this line (Ares and Poseidon) correspond to the first and last of the triad of gods in the E/B god-list (Ba$^{c}$al-Šamem and Ba$^{c}$al-Ṣaphon), the question naturally arises as to whether the second gods in the lists (Triton and Ba$^{c}$al-Malage) might be related. There is an added difficulty here: although the gods Ba$^{c}$al-Šamem and Ba$^{c}$al-Ṣaphon are well known from Phoenician and Punic inscriptions, Ba$^{c}$al-Malage is attested only here. Like the other two divine names in this line of E/B, Ba$^{c}$al-Malage is probably an epithet of a deity--but which deity? But for the moment we should concern ourselves with the prior question, who is Triton here?

Dussaud has identified Triton with the Phoenician god Kušor.[401] Walbank, however, is skeptical of this identification:

> The equation Triton-Kousor is ill-based, since whatever the resemblances between Kousor as a sea-god . . . and Triton--frankly, they are few--in the Ras Shamra myth Kousor assists Baal against Yam, whereas Triton, though perhaps originally an earlier sea-god than Poseidon, is represented as his son and is never in conflict with him. . . .[402]

But again one should bear in mind that in general the equivalence of a Greek with a Punic deity is based upon one or at most several common traits or functions; one should not expect two gods to correspond in every detail. We have seen, for example, that the equivalence of Zeus and Ba$^{c}$al-Ḥamon in our list appears to be based on only one common trait: supremacy in the pantheon.

At Ugarit Kušor (Ugaritic *Kôtar*) is generally presented under his aspect of craftsman or artisan.[403] It is perhaps this aspect that explains his connection with Memphis (Ugaritic *hkpt* < Egyptian *ht k3 ptḥ*, "sanctuary of Ptaḥ"), the city of Ptaḥ, the Egyptian craftsman-god; Mochos of Sidon calls Kušor the first "Opener," undoubtedly a pun on the name Ptaḥ and the West Semitic verb *ptḥ*, "to open."[404] Later he was equated with the Greek smith-god, Hephaestus.[405] But even at Ugarit Kušor's connections with the sea are intimated by the fact that he is called *bn ym*, "the son of Yamm" (*UT* 4[51].7.15-16). This connection, however, is much stronger in Sakkunyaton. Here Kušor is pictured, it is true, as one who "concerned himself with sayings [*logous*], incantations, and oracles";[406] but the account goes on to say, "He [i.e., Kušor] is Hephaestus; he invented the fishhook, the lure, the fishing-line, and the light boat [*schedia*]; he was the first of all men to sail a ship."[407] Apparently the marine associations of Kušor in Phoenicia were more prominent--or perhaps had become more prominent (cf. the case of Ba$^c$al-Ṣaphon vis-à-vis Hadad)--than at Ugarit.

If Dussaud's identification of Triton with Kušor is correct, the only aspect common to the two would appear to be that both are intimately associated with the sea. Precisely why Triton was chosen rather than, say, Pontos is not clear; the choice could have been arbitrary to some extent.

Assuming for the moment that Dussaud is correct, we must now determine whether there is any relation between Triton/Kušor and Ba$^c$al-Malage. A. García y Bellido notes that on certain ancient coins from what is now Málaga in Spain there appears a god "carrying tongs as his [specifying] attribute";[408] now tongs are one of the common attributes of Hephaestus.[409] García y Bellido claims, "It is beyond doubt that this Punic deity, whatever his identity, became a tutelary deity of Malaca";[410] he implies that the god in question is in fact Kušor, since he places this whole discussion under the sub-heading *Chusor*. It is not impossible that the element *malagê* (*-ma-la-ge-e*, a reading that S. Langdon has termed "certain")[411] is connected with the place-name Malaca (Greek *Malaka*).[412] This may explain the puzzling[413] long vowel at the end (*-ê*): it could be a gentilic suffix.[414] The fact that the Assyrian text reads the consonant *g* rather than *k*--the sign *gi/e*, which appears here, had no other phonetic value during this period[415]--is admittedly a minor problem; but the presence of the voiced rather than the voiceless stop may indicate no more than a slight inaccuracy on the part of the Assyrian drafters of the treaty.

Yet another connection between Ba$^c$al-Malage and Kušor may be inferred from a suggestion made recently by S. Moscati. Moscati posits an identification between the Ba$^c$al-Malage of E/B and Zeus Meilichios mentioned in Sakkunyaton.[416] But what he does not mention is that Sakkunyaton seems to equate Zeus Meilichios with Kušor: "He [i.e., Kušor][417] was also called Zeus Meilichios."[418] Now *m(e)ilichios*, meaning "gentle," was an epithet of Zeus at Athens and in Boeotia and was also applied to other Greek divinities.[419] But obviously here the name refers to a Phoenician rather than a Greek god; for it is difficult to see what Zeus and Kušor had in common beyond the fact that they were both gods. Hence we might expect "Zeus" here to represent the element "Ba^c^al" as in *Zeus Ammōn* ( = Ba^c^al-Ḥamon),[420] *Zeus Ouranios* ( = Ba^c^al-Šamêm),[421] or *Zeus Kasios* ( = Ba^c^al-Ṣaphon). In the opinion of a number of scholars the element *m(e)ilichios* here actually reflects an underlying Phoenician word; it is generally thought that this is related to the root *mlḥ*, whence Hebrew *mallāḥ*, "sailor."[422] However, it is Ba^c^al-Ṣaphon, as we have seen, who was the patron of sailors. Others have suggested a connection with *mlk*. S. Langdon thinks of *mâlak* (root actually *l'k*) as in "Mâlak-Bêl, the sun-god of Palmyra."[423] But a sun-god does not fit here. We would postulate a connection with the place-name Malaca, which also appears as Malacha.[424] The ending *-ios* might point to an adjective derived from a place-name (cf. *Korinthos, Korinthios*, etc.), like the apparently gentilic ending of *malagê*. Thus a putative original *Zeus *Malachios* ("Ba^c^al of Malaca") could have become *Zeus Meilichios*, since the latter would be more familiar to a Greek copyist, and since in any case, *malachios* (the name of a certain type of fish)[425] may have seemed inappropriate to a Greek as the epithet of a deity. The underlying Phoenician name would have been something like **Ba^c^al-Malakī*, with the Phoenician gentilic suffix.

In summary, we postulate (with Moscati) an identification of Ba^c^al-Malage with Zeus Meilichios ( = Kušor) of Sakkunyaton on the one hand, and (with Dussaud) the Triton of our treaty with Kušor on the other. If this is correct the deity here called Triton appears--as one might expect by now--also in the Tyrian god-list in E/B, under the name Ba^c^al-Malage. We noted above that the curse that follows the names of the three gods in E/B indicates that each of them is associated with storm and/or the sea. While there is no evidence that Kušor was ever regarded as a storm-god, he was definitely associated with the sea in Phoenicia (according to Sakkunyaton), as was Triton in Greek mythology.

If our interpretation of this section is correct, this last line of high gods in H/P corresponds exactly--even as to sequence--to the triad of Tyrian "Ba^c^als" in the E/B list of Phoenician gods; and furthermore, every one of the Tyrian deities named in E/B is named in the high gods section of H/P.

## B. Gods Paralleled in the Hittite Treaty Tradition

The next three lines of the list of divine witnesses clearly differ from the preceding three. The latter listed the names of the high gods of the official pantheon of Carthage; the three lines now under discussion, however, contain no proper names and name no high gods. The last two lines present divinized natural phenomena, such as we have encountered in the Hittite treaties and Sf1, whereas the import of the first line is not readily apparent. But the irrefutable presence of the natural phenomena here, in a section after the high gods and before the summary phrases and the statement of divine attestation, strongly suggests a parallel to Sf1, which follows exactly the same arrangement. In turn, the god-lists both of H/P and Sf1 show influence from the Hittite treaty god-lists.

### 1. ἐναντίον θεῶν τῶν συστρατευομένων

The first problem one faces in interpreting this phrase is syntatical. Is the genitive *tōn systrateuomenōn* to be construed as in apposition to *theōn* or as a possessive genitive? In the first case the phrase would be rendered "the gods who take the field with (us)," and in the second, "the gods *of those who* take the field with (us)." By far the first is the more commonly accepted understanding,[426] although the second is not without endorsement.[427]

But the first interpretation presents a difficulty. If Walbank is correct in maintaining that "these *theoi* accompany the Punic army,"[428] how do they differ from *theōn pantōn tōn kata strateian* in the next section? Walbank frankly admits this difficulty[429] but offers no suggestions for a solution. One cannot lightly assume a redundancy in so carefully structured a god-list as that of H/P. And the gods in question can hardly be those of Carthage or Macedonia or of the army, since these three categories are specifically listed in the next section.

To get around these difficulties one must opt for the second interpretation of the phrase: "the gods *of those who* take the field with (us)." The gods would not be those of the Punic or Macedonian troops but rather those of "the various nationalities represented in Hannibal's camp[, who] would have their separate gods,"[430] that is, the gods of the various allies or mercenary contingents.

This understanding of the expression in question also fits well with the Hittite treaty tradition. We have seen that in Sf1 the section containing the high gods is followed by one line containing *ʾl wʿlyn*, which corresponds to one of the Hittite olden gods, and three pairs of divinized natural phenomena. But it is unlikely that the line under discussion in H/P has any connection with the olden gods. Again comparison with the Hittite treaties may prove helpful. Just before the summary phrases, the olden gods, and the deified natural elements comes what we have called the "Gods of Mercenaries." This sub-section always contains only one entry, namely, "the gods (of the?) Lulaḫḫi, the gods (of the?) Ḫapiri." One should note too that these gods never qualified by the word "all,"[431] whereas the summary phrases (referring to the gods of Hatti and the vassal territory) amost always contain this word. This usage is strikingly parallel to the H/P list, where *pantōn* is not used with *theōn tōn systrateuomenōn* but does occur in each line of the summary statements in the next section. These considerations seem to indicate that this line of the Carthaginian treaty corresponds to our Section VII B of the Hittite treaties.

### 2. [ἐναντίον] Ἡλίου καὶ Σελήνης καὶ Γῆς

The surviving manuscripts of our treaty do not contain the word *enantion* here. But it must have stood in the text originally, as may be seen from the following arguments. We have noted that the god-list is tripartite:[432] the first section consists of three lines, each containing three divine names; the third section likewise consists of three lines, each beginning with the phrase "all the gods of"; in the middle section the last two lines each contain three deities, as in the first section. This indicates that *theōn tōn systrateuomenōn* and *Hēliou kai Selēnēs kai Gēs* are to be regarded as two separate lines; only in this way does the central section contain three lines like the other sections. Moreover, if these two lines were read as one, the resulting unit would contain more than three deities, which would conflict with the symmetry of the first two sections. It will also be remembered that in the god-list from Sf1, the list most parallel in structure to the H/P god-list, each line (which consists of no more than a pair of deities) is preceded by *qdm*, "in the presence of." The first line of this section is the only one in the first two sections that does not contain the names of three deities; thus a copyist unaware of the structure might easily have sensed something unusual about this line (the shortest one in the god-list) and joined it with the following line, inadvertently omitting the original *enantion* in the process. The *kai* before *Hēliou* would have been inserted as a connective sometime after the loss of *enantion*. Hence in the lemma above we restore *enantion* and omit the first *kai*.

Perhaps more than any other part of the list these two lines exhibit a sense of symmetry and balance. For this reason it is appropriate that we point out some of the more readily apparent symmetrical features common to both lines. Since our concern here is primarily with the second line of this section, however, the discussion of some of the stylistic features of the third line must be deferred.

First, as is obvious, each of these lines contains a "triad" of divinized natural elements. Furthermore, the second line provides a balanced contrast to the third: each of the natural elements in the second line is in the singular, whereas each in the third is in the plural. Lastly, the second line lists only celestial/terrestrial bodies, whereas the third lists only--as we shall see--aqueous elements.

In Sf1 the arrangement of the divinized elements is in natural pairs: heaven and earth, abyss and springs,[433] day and night. These "natural" or "fixed pairs" of complementary words are frequently found in the Semitic languages, including Phoenician and Punic. Thus the arrangement of the elements in threes is quite artificial; it merely serves to continue the triadic pattern established in the first section and reflected in the tripartite division of the list as a whole. Yet the internal structure of the line is even closer than might be immediately perceived to that of the first line of the god-list, which sets the pattern throughout. The first two names form a "natural pair" as god and consort (Baᶜal-Ḥamon and Tanit), while the third (Rešep) in turn is "paired" with these as child to parents. So too in the present line "Sun and Moon" clearly form a fixed or "natural pair" in West Semitic,[434] including Phoenician.[435] The expected complement of "Earth," however, is neither "Sun" nor "Moon" but "Heaven": "Heaven and Earth" is one of the most common fixed pairs in ancient Semitic literature.[436] This pair, in fact, is listed first among the divinized natural phenomena not only in Sf1 but also in some Hittite treaties[437] and, significantly, in the only Phoenician inscription closely connected with the treaty tradition, the Arslan Tash amulet (*KAI* 27).[438] The pair also figures in the "pantheon list(s)" from Ugarit.[439] So too in the first triad of natural elements in the H/P list "Earth" is "paired" with the natural pair "Sun and Moon" as the celestial with the terrestrial. In other words, in order to accommodate the triadic structure the expected pair "Heaven and Earth" has been artificially expanded into "Sun-and-Moon and Earth"; yet it is difficult to deny that this triad stands in direct association with "Heaven and Earth" in other lists of divine witnesses. One can only admire this sense of balance; the patterns of the ancient treaty tradition were not disregarded but were nonetheless modified to accommodate an apparently more recent development, the grouping of the deities in threes.

### 3. ἐναντίον ποταμῶν καὶ λιμνῶν καὶ ὑδάτων

This last line of the second section also lists three divinized natural elements; all are in the plural and at least the first and last have in common an "aqueous" character. There is a textual problem with the second member. The manuscript itself reads *daimonōn* ("deities/demons"); Causabon suggested *leimōnōn* ("meadows"); Gronovius, *limenōn* ("harbors"); and Reiske, *limnōn* ("lakes").[440] Walbank favors the reading *limnōn*,[441] which has also been defended recently by M. Weinfeld.[442] This reading is certainly to be preferred for a number of reasons. First of all, the other two members mentioned are clearly "watery" elements; one would thus expect the second member to be also. Secondly, after the primordial pair "Heaven and Earth" (corresponding to "Sun, Moon, and Earth" in H/P) in many treaty and/or cosmogonic lists comes the listing of a pair of watery elements. For example note "Abyss and Springs" immediately after "Heaven and Earth" in Sf1; *thwm* ("deep") // *hmym* ("the waters") in Gen 1:2; *ymym* ("seas") // *thmwt* ("deeps") just after *šmym* // *ʾrṣ* in Ps 135:6.[443] In the Akkadian *Erra* epic (1.136), also in a "cosmic" context, *nagbu* ("springs") // *mīlu* ("high water") follow immediately after AN$^{e}$ *u* KI$^{tim}$ ("heaven and earth").[444] In the previous line of our list, the original pair was expanded to three without destroying the opposition of the heavenly (Sun and Moon) to the earthly (Earth); so too here we would expect an original pair of parallel aqueous elements to be expanded to three. The second term in this triad, then, could hardly be "demons" or "meadows."

Of course, theoretically *limenōn* ("harbors") would seem to fit this pattern as well. But the natural elements invoked in treaties have another common characteristic lacking in "harbors." All of these elements --mountains, rivers, springs, heaven, earth, winds, the sea--have a primordial aspect associated with the cosmogony. For this reason a number of them (especially rivers, winds, sea) also possess a chaotic aspect, the potential for uncontrollable destruction. At the same time, however, some of these same elements have a life-producing or fertilizing character (especially springs, clouds, rivers). The term "harbors" possesses none of these characteristics, which explains why it is never found in these cosmogonic lists. The "watery" elements that do occur in such lists and possess one or several of the characteristics just noted are usually rivers, springs, the deep, and the sea.

One might object at this point that "lakes" never appears in the lists of "cosmogonic" aqueous elements either. This is quite correct and leads us to question the meaning of *limnōn* here. The Greek word (singular, *limnē*) denotes a "*pool of standing water* left by the sea or a river"[445] in classical Greek; in the Greek of the Hellenistic period it can also mean "lake."[446] This prosaic sense hardly seems to fit the nature of the elements found in the cosmogonic lists. However, the word is also used in Homer and especially in tragic and lyric poetry to mean the *sea*.[447] Now the sea is one of the most obvious and important cosmogonic elements; it is listed therefore in the great majority of cosmogonic lists, those within treaties as well as elsewhere. We have already noted the frequent association of the sea with "Heaven and Earth" in such lists from the OT. "The great sea"[448] appears among the natural elements invoked in the great majority of Hittite treaties; "Sea" (*ym*) also occurs in the "pantheon list(s)" from Ugarit.[449]

The precise meaning of *limnōn* in the god-list may also be sought from another direction: what Punic word is being translated here? The word in question is most probably *ymm*. For one thing, the word *yamm* in West Semitic denotes not only the "sea" as we generally think of it but any large body of water, including large rivers and lakes.[450] Hence inland "seas" or "lakes" are always designated as *ym* in Hebrew: for example, *ym hmlḥ*, "the Dead Sea" (lit., "the Sea of Salt"). Interestingly the "Sea of Galilee" or "Lake Gennesaret" (Hebrew *ym knrt*: Num 34:11; Josh 13:27) is translated *hē limnē Gennēsaret* in Lk 5:1, though it is usually called *thalassa* elsewhere.[451] Josephus calls it *limnē Gennēsar*[452] or simply *hē limnē*.[453] The word *ym* appears also in Phoenician, where only a few occurrences are attested;[454] there it means "sea." But given the fact that this word has a wide semantic range in West Semitic, it would be captious to deny that it had the same range in Phoenician-Punic as well.

There is a yet more compelling argument for holding that Punic *ymm* stands behind *limnōn* here. As Baᶜal-Ḥamon and Tanit on the one hand and

Sun and Moon on the other (also probably the "protective deity of the Carthaginians" [ᶜAštart] and Melqart) form natural pairs, so "river(s)" and "sea(s)" form such a pair in West Semitic literature. This fact is abundantly attested both in Ugaritic texts and the OT. In the Ugaritic Baᶜal cycle we find *zbl ym* ("Prince Sea") // *ṯpṭ nhr* ("Judge River").[455] Both names denote the same individual. Here "Sea/River" refers to the deified, chaotic element subdued by Baᶜal; the same is true of Tiamat (also meaning "Sea") vanquished by Marduk in *Enūma Eliš* and of Sea conquered by Yahweh in the OT.[456] Moreover, *ym(ym)* is frequently parallel to *nhr(wt)*, "river(s)," in Hebrew poetry.[457] What is more, *ym(ym)* is often found in parallelism with *mym (rbym)*, "(mighty) waters."[458] Thus *ymm* would be expected in the Punic original[459] not only because it alone would form a fixed pair with "rivers" but also because it forms such a pair with "waters" (*hydatōn* in H/P) as well. These considerations, then, strongly suggest the presence of Punic *ymm* in the original text of the god-list. The Greek *limnōn*--if it means "lakes" here--may be explained as a plausible if somewhat inaccurate translation of the West Semitic *ymm*.

The fact that *ym(ym)* in Hebrew poetry is often parallel to *mym (rbym)* may help us to understand the exact nuance of *hydatōn* in our text. Here the reference is not simply to the chemical element $H_2O$ or to any body of water[460] but to "waters" as a cosmic force. This is why the parallelism is usually between *ym(ym)* and *mym rbym*, "the *mighty* waters.[461] Even without the qualification *rbym* it is apparent that *mym* has this connotation, as can be seen from passages such as Pss 29:3 and 124:4-5, where *mym* and *mym rbym* are used interchangeably. Now "the mighty waters" are "the chaotic, disorderly, insurgent elements which must be controlled."[462] In the ancient Semitic conception of the world these surrounded the earth, posing the constant threat of a return to primeval chaos. Apparently too this phrase can refer to chaotic or primordial waters of any kind; this explains its occasional parallelism with *nhrwt*, "rivers" (Cant 8:7) or *thwm(wt)*, "the deep(s)" (Ezek 26:19) as well as *ym(ym)*. This understanding of "waters" may best fit the cosmic or cosmogonic context of the deified elements in treaty god-lists. Moreover, only when conceived in this way are river(s), sea(s), and waters ever personified, as they are in mythological texts. This is, incidentally, further evidence that the interpretation "lakes" for *limnōn* has no place in such a context and hence cannot be correct.

But there is another possible interpretation of *hydatōn* here. As noted above, various cosmic "watery" elements occur regularly in the Hittite treaty lists of deified natural elements. There are almost always three: "rivers" (ÍD.MEŠ),[463] "the great sea" (A.AB.BA GAL/*šalliš arunaš*), and "springs" (TÚL.MEŠ). The precise correspondence of the first two with *potamōn* and *limnōn* (except that "sea" is not plural in the Hittite treaties) suggests that *hydatōn* may correspond to TÚL.MEŠ. In Hebrew texts *mym rbym* sometimes appears with the emphasis on the primordial rather than the chaotic aspect of "waters." In such cases the phrase refers "to the waters of the deep beneath the earth as the source of fertilizing waters,"[464] as in Ezek 17:5; 31:7; and Num 24:5-7. Thus *mym (rbym)* is found in parallelism with *thwm(wt)*, "the deep(s)," the source of all groundwater in wells and springs.[465] Thus it is possible that *hydatōn* might refer to the fertilizing waters that come from below the earth and, therefore, be a reflection of TÚL.MEŠ in the treaty tradition (cf. *m$^c$ynn*, "springs," in Sf1). Hence it is difficult to decide which nuance *hydatōn* has here: that of "mighty waters" (the chaotic aspect), as suggested by the Hebrew parallelism with "sea(s)"; or "groundwaters" (the primordial, fertilizing aspect), as suggested by the presence of TÚL.MEŠ and *m$^c$ynn* in the treaty tradition.

### C. Summary Phrases

1. ἐναντίον πάντων θεῶν ὅσοι κατέχουσι Καρχηδόνα
   ἐναντίον θεῶν πάντων ὅσοι Μακεδονίαν
   καὶ τὴν ἄλλην Ἑλλάδα κατέχουσιν

The structural parallels between these lines and the summary phrases from other ancient treaties were touched upon in Chapter III. The use of the term "all" in each line of this section recalls the presence of this word in the corresponding lines of the Hittite treaty god-lists (*ḫūmanteš/ḫūmanduš*) as well as of Sf1 (*kl*). These two lines of H/P also resemble Sf1 in the position of the summary phrases: in both treaties these phrases come at the end (although in Sf1 a similar expression occurs within the canonical oath-gods section), whereas they appear before the olden gods and natural elements in the Hittite treaties. But in another respect the H/P list follows more closely the Hittite pattern: whereas in Sf1 the names of the two contracting nations apparently are mentioned in a single statement, "All the g[ods of KTK and Arpad . . .]," in the Hittite treaties and in H/P there are separate lines listing the contracting parties. In both cases, of course, the country of the party drafting the treaty is named first:

DINGIR.MEŠ LÚ.MEŠ DINGIR.MEŠ SAL.MEŠ
*gap-pa-šu-nu ša* KUR $^{uru}$*Ḫa-at-ti*
DINGIR.MEŠ LÚ.MEŠ DINGIR.MEŠ SAL.MEŠ
*gap-pa-šu-nu ša* KUR $^{uru}$*Ki-iz-zu-wa-at-ni*
DINGIR.MEŠ LÚ.MEŠ DINGIR.MEŠ SAL.MEŠ
*gap-pa-šu-nu ša* KUR $^{uru}$*Nu-ḫaš-ši*.[466]

All the gods (and) goddesses of Hatti,
All the gods (and) goddesses of Kizzuwatni,
All the gods (and) goddesses of Nuḫašši.

Compare the present lines of our text:

All the gods who dwell in Carthage,
All the gods who dwell in Macedonia and the rest of Greece.

Note that we have rendered the phrase *hosoi katechousi(n)* as "who *dwell* in." Almost all other commentators translate "who *possess*" Carthage, etc."[467] The one exception is E. Meyer, whose translation agrees with ours.[468] The verb *katechein* can have either nuance, it is true,[469] and the meaning "to inhabit/dwell in" appears to be attested for this word largely in lyric poetry.[470] But to speak of gods who "possess" a certain area is not Semitic idiom; to speak of gods who "dwell" in a place is. "All the gods who dwell in Carthage" is semantically equivalent in Semitic phraseology to "All the gods of Carthage." This may be seen from geographical epithets of the gods, especially in Akkadian:

| | |
|---|---|
| $^{d}$*Ištar ša* $^{uru}$*Ninua* | Ištar *of* Nineveh |
| $^{d}$*Ištar āšibat* $^{uru}$*Ninua* | Ištar, *who dwells in* Nineveh |
| $^{d}$*Ištar bēlet* $^{uru}$*Ninua*[471] | Ištar, *Lady of* Nineveh. |

The Hittite treaties in Akkadian make use of the first construction, where the divine name is connected to the place-name by means of *ša*; H/P, however, uses the second,[472] the participle *yšb* in Punic or perhaps the relative clause *ʾš yšb*. The participial construction is well attested in West Semitic.[473]

Here again the translators of this treaty seem to have avoided the most common or expected equivalent of the Punic word and chosen what may perhaps be called a "poetic" term. For *yšb* we would have expected *(kat)-oikein*, the ordinary word in Greek for "dwell" or "inhabit," just as for *ymm* ("seas") in the preceding section one would have expected *thalassōn* rather than *limnōn*.

2. ἐναντίον θεῶν πάντων τῶν κατὰ στρατείαν

There are two points of interpretation to be discussed in this line. The more important is the meaning of *strateia*. This word commonly means *"campaign, warfare,"* etc.[474] However, Liddell and Scott point out that "*stratia* ["army"] is a constant v[aria] l[ectio] . . ." of this word, and moreover that *strateia* itself can signify *"army, expeditionary force,"* although this usage is "very rare."[475] Because *ei* and *i* were homophonous by the time of this treaty,[476] either *strateia* or *stratia* could be the original reading. However, we see no reason to emend the line as it now reads.

In Chapter III we noted[477] that one Hittite treaty god-list (F6 I.52-53) contains the expression [DIN]GIR.MEŠ KARAŠ *ḫu-u-ma-an-te-eš* ( = Hittite *tuzziyaš šiuneš ḫūmanteš*), "all the gods of the army/camp." This phrase comes at the end of a section of war-gods and thus is a true "summary phrase," which explains the use of "all" (*ḫūmanteš*). The Sumerogram KARAŠ (or KI.KAL.BAD) corresponds in Akkadian to *karāšu*, which means "camp" or "expeditionary force; campaigning army."[478] The Hittite *tuzzi-*, which stands behind KARAŠ, also has both of these connotations.[479] Significantly, West Semitic *mḥnt* shares the same semantic range, in Phoenician-Punic[480] and Hebrew.[481] In all likelihood, then, *strateia* here is a translation of the Punic *mḥnt*. Hence it is to be translated "army," since neither *strateia* nor *stratia* has the significance "camp."

A second point concerns the syntax of *tōn kata strateian*. Theoretically, this could represent the genitive plural of an expression *ta kata strateian*, "the things that pertain to the army."[482] But the resulting translation would give the improbable expression, "all the gods of army affairs"! It is far more likely that the *kata* phrase represents a periphrasis of a genitive noun, a construction well attested from Plato on.[483] For example, we read elsewhere in Polybius (3.113.1): *hē kata ton hēlion anatolē* ( = *hē tou hēliou anatolē*), "the rising of the sun." Every translation of this line of the god-list reflects this interpretation of the syntax. Hence the line is to be rendered, "(In the presence of) all the gods of the army"--the verbatim equivalent of the Hittite *tuzziyaš šiuneš ḫumanteš*. It is important to note that among the treaties of the ancient Near East this particular summary phrase occurs only in H/P and in one Hittite treaty.

But why does this expression occur at this point? Unlike the case of F6 the phrase in H/P does not summarize a list of deities who are specifically war-gods. Once again the answer seems to lie in the structure of the list of divine witnesses. Considering the pattern of the Hittite summary phrases, which the lines of the present section clearly parallel, we might expect only two lines here, since only two nations are contracting parties in this treaty--Carthage and Macedonia. But to have only two lines in this section would destroy the threefold pattern set up in the preceding sections. Thus it is conceivable that the third summary phrase was added to balance the other two sections. Because of its position in the final section of the list, it is perhaps to be understood as part of a general summary: the reference is probably to all the gods who are thought to fight on the side of the Carthaginian forces.

## D. Statement of Divine Attestation

ὅσοι τινὲς ἐφεστήκασιν ἐπὶ τοῦδε τοῦ ὅρκου

The position of this statement of divine attestation to the treaty accords precisely with that of Sf1, in which the statement/invocation comes at the end of the god-list. In the Hittite treaties these invocations are placed either just before or just after the list. It must be emphasized that in all these treaties the statement concerns all the gods who have been invoked. Although it might seem belaboring the obvious to point out this fact, it is necessary to do so in order to understand the function of the phrase. Several commentators restrict the antecedent of *hosoi tines* to the preceding line, "all the gods of the army."[484] But in light of what has just been said the statement of invocation must have a more general reference. Thus a translation such as "as many as"[485] for *hosoi tines* is to be preferred.

Next we must try to determine more precisely what this clause means. Exactly what are the divine witnesses said to do here? Proposed translations include "watch over,"[486] "preside over,"[487] "are witnesses to,"[488] "stand by at."[489] Here again the treaty tradition can be helpful. Some of the Hittite treaties in Akkadian also employ the verb "to stand" (*izuzzu*) in the summons to the gods.[490] It is significant that this particular expression within the invocation is found--again--only in the Hittite treaties (in Akkadian). But specifically what does *izuzzu* signify in this context? Some of the treaties contain variants which provide a clue. For example, in W2 Rev 10-11 we read:

> *i-na* ŠÀ *ri*[*ik-si an-ni-i* . . . DINGIR.
> MEŠ . . . *ni-il-ta-as-sí*] *a-na a-ḫa-m*[*iš*
> *li-iz-zi-iz-zu li-el-te-mu-u ù lu-u*
> *ši-bu-tù*][491]

> At (the conclusion of) [this] trea[ty . . .
> we have called (upon) . . . the gods];
> [let them stand] *togethe*[*r*, let them hear,
> and let them be witnesses].

The fact that in this text, unlike the others, the gods are asked to stand together points to another unique expression in a Hittite treaty in Akkadian in which we read:

> [DINGIR.]MEŠ *li-ip-ḫu-ru l*[*i-iš-*
> *te-mu-ú ù lu-ú še-bu-tu*[492]

> Let [the gods] *assemble*, l[et them hear,
> and let them be witnesses].

It is interesting that, although Hittite treaties in Hittite outnumber the copies in Akkadian, no text has been found to date that contains a Hittite equivalent of *lizzizzū*, "let them stand." However, the summons to "stand *together*" or "*assemble*" reflects a word commonly found in the treaties in Hittite, *tuliya*.[493] This is the dative-locative form of *tuli-*, usually translated "assembly" (*Ratsversammlung*).[494] It occurs in the phrase *tu-li-ya ḫal-zi-ú-en*,[495] "we have called (the gods) to assembly"; and *tu-li-ya ḫal-zi-ya-an-te-eš*,[496] "(the gods) who are called to assembly." It seems likely, then, that the function of the term *tuliya*, which has no direct equivalent in the treaties in Akkadian, is reflected by *ana aḫāmiš lizzizzū* and *lipḫurū*, "let them *stand together/assemble*."

But the most striking parallel in this line to the summons in the Hittite treaties is *epi toude tou horkou*. More than anything else it is the interpretation of this phrase that holds the clue to the precise meaning of the whole sentence. Unfortunately *epi* (with the genitive) is very ambiguous. But the corresponding phrase in the Hittite treaties is not so ambiguous:

> *[nu k]a-a-ša ke-e-da-ni li-in-ki-ya 1 LI-IM*
> DINGIR.MEŠ *tu-l[i-ya ḫal-zi-ú-en]*[497]
>
> [And be]hold, [we have called] the thousand gods to as[sembly] *at* (the conclusion of) *this treaty*.

Here the phrase *kēdani linkiya*, "*at* (the conclusion of) *this treaty*" (lit., "oath"), is in the dative-locative case. That the locative syntax is intended is quite clear from the Akkadian texts, which read *ina* ŠÀ/*libbi*, "in."[498] Note, for example, W2 Rev 10-11 (p. 97) and the following passage:

> *i-na lìb-bi a-ma-te*.MEŠ *an-nu-ti ša ri-ik-si*
> *li-iz-zi-iz-zu*[499]
>
> Let them stand by/be present *at* (the utterance of) these *treaty*-stipulations.

This interpretation of the call to divine attestation in Hittite treaties and in H/P is confirmed by the same use of the verb "to stand" in Babylonian *kudurrus*. The word occurs in the section listing the witnesses to the document. The form of the statement of attestation is almost identical to that found in the Hittite treaties:

| | |
|---|---|
| *i-na ka-nak* | *i-na lìb-bi* |
| *ṭup-pi šu-a-tu* | *rik-sí an-ni-i* |
| $PN_1$, $PN_2$, $PN_3$, | DINGIR.MEŠ *ša pu-uz-ri* |

| | |
|---|---|
| $PN_4$, $PN_5$, etc. | *ù* DINGIR.MEŠ *ša* EN |
| [witnesses] | *ma-mi-ti ni-il-ta-as-sí* |
| *iz-za-az-zu*[500] | *l[i-l]z-zi-iz-zu* |
| | *li-il*$_5$*-te-mu-u ù* |
| | *lu-ú ši-bu-tù*[501] |

| | |
|---|---|
| At the sealing of | At (the conclusion of)[502] |
| this tablet | this treaty we have summoned |
| $PN_1$, $PN_2$, $PN_3$, | the hidden gods and |
| $PN_4$, $PN_5$, etc. | the gods who are |
| [witnesses] | lords of the oath; |
| are *standing by/present.* | let them *stand by/be present*, |
| | let them hear, and |
| | let them be witnesses. |

It is obvious that "standing by" in both texts refers to being present as witnesses to the document in question. The verb *ephistēmi* can also mean "to stand by, be present,"[503] which is certainly the meaning of *ephestē-kasin* in H/P.

Thus *ephestēkasin epi toude tou horkou* is, except for tense and mood, exactly equivalent to Akkadian *ina libbi riksi annî . . . lizzizzū* and *epi toude tou horkou* is equivalent to Hittite *kēdani linkiya*. Of the proposed translations of this line, then, the most accurate is "stand by at" and "are witnesses to" (M. Weinfeld, E. S. Shuckburgh). We would translate, "who are present (as witnesses) at (the conclusion of) this treaty."[504]

## Chapter V

## CONCLUSIONS

In the course of this work we have had as our main purpose to determine what light could be shed on the structure and meaning of the H/P god-list from an analysis of similar lists in ancient Near Eastern treaties. The evidence presented leads to the conclusion that a close connection exists between the list of divine witnesses in H/P and the god-lists from other ancient Near Eastern treaties in Hittite, Akkadian, and Aramaic. Since some of these date over a thousand years before the time of Hannibal, the detailed correspondence in structure and phraseology is striking. But there are numerous examples of the conservatism of legal and religious traditions throughout history; and since a treaty god-list draws from both, the doubly reinforced conservatism or adherence to ancient structures and formulas is almost to be expected. The list of divine witnesses in H/P, the latest treaty of this type known from the ancient world, stands as a testimony to the longevity and endurance of these traditions.

The link with the past as seen in this treaty is in a sense twofold: a link with the divine witnesses of the treaty tradition of first-millennium B.C. Tyre, the mother-city, and through it a link with the god-lists of the Hittite treaties. What we know of Tyrian treaty god-lists is based on a single document (E/B) and concerns only the "high gods." Yet it appears that all of these deities, the gods of the official pantheon of Tyre in the seventh century B.C., are invoked in the first section of the H/P list. Moreover, some of the gods are listed in the same sequence in both documents.

Since the choice of which high gods were to be invoked in the first section of the list was fixed by a tradition stretching back at least five hundred years, the god-list itself shows very little of the imprint of Hannibal. The undeniable connections of form and content between this list and those in other ancient Near Eastern treaties, where the god-list always consists of deities from the official pantheon, as well as the correspondence between the H/P god-list and the E/B list of foreign gods contradicts Picard's theory that the gods in H/P are the dynastic deities of the Barcides. Undoubtedly if this treaty had been made under the auspices of any other Carthaginian figure--Barcide or not--during this period, the god-list would have differed little if at all.

The section of the high gods contains no surprises. Polybius notes that in the treaty of Carthage with Rome between 280 and 278 B.C. the Carthaginians swore "by their ancestral gods" (*Histories* 3.25). In our text too there is no doubt that all the gods invoked are native Punic deities, though their names are translated by the names of the Greek gods. All the major deities of Carthage are included. The prevailing belief that in this god-list we find the official pantheon of Carthage appears fully justified.

One interesting departure from the structure of earlier god-lists is the dominant role of the triadic pattern; in the older texts we occasionally note a dyadic structure but never a triadic one. It is not possible at this point to state how extensively Carthaginian deities were grouped in threes or how rigid these triads were; but their presence in an official state list of gods indicates that they were clearly not random patterns.

The careful attention to symmetry and balance evident in every section of the list is also noteworthy. Apparently on the model of the divine triad in the first line, the list as a whole divides into three clearly distinct sections (exclusive of the statement of divine attestation). Each of these, in turn, consists of three lines. And in the first and second sections, each line (except one) consists of three entires. In the case of the natural phenomena, fixed pairs of elements (well attested in West Semitic literature) have been ingeniously expanded to accommodate the triadic pattern, while the sense of the original pairs has been retained.

But the structural patterns in this list reach back even further in time to the Hittite state treaties of the second millennium B.C. With the exception of the position of the summary phrases (which follow the Syro-Mesopotamian model), the H/P god-list reflects the Hittite arrangement precisely. Even within each of the sections there are further possible reflections of the Hittite structure. The three lines in the high gods section appear to begin with the same divine types (i.e., supreme deity, protective deity, war-deity) found in the first three sub-divisions of the corresponding section of the treaties from Hatti. In the second section the first line refers most probably to the gods of the mercenary contingents, while the second two contain divinized natural elements (a feature entirely lacking in Mesopotamian treaties); even the order of the elements is approximately the same as in a number of the Hittite treaties. The "gods of those who take the field with (us)" also follows the Hittite pattern.

One can even point to a number of verbatim parallels between the two treaty formularies. The anaphoric use of *enantion* throughout the list apparently goes back to a Syrian form (Sf1) and ultimately to the Mesopotamian tradition (cf. *ina* IGI, "in the presence of," in several Assyrian treaties). But aside from this the verbal parallels point in the direction of Hatti. Only in the Hittite treaty tradition do we find the invocation of a "protective deity" (Greek *daimōn* = Hittite LAMMA/*annari-*) or of "all the gods of the army" (Greek *theōn pantōn tōn kata strateian* = Hittite *tuzziyaš šiuneš ḫūmanteš*). Only in this tradition are the gods asked to "be present" (Greek *ephestēkasin* = Akkadian *lizzizzū*) as witnesses "at (the conclusion of) this treaty" (Greek *epi toude tou horkou* = Hittite *kēdani linkiya* = Akkadian *ina libbi riksi annî*). And only in the Hittite tradition does one find "all the gods" of each contracting party listed in separate lines.

The god-list reveals a close relationship not only between the treaty traditions of Carthage and Tyre but also between the official pantheons of these two cities (Appendix I, Chart 17). The supreme status of Ba$^{c}$al-Ḥamon and Tanit at Carthage occasions no surprise, although the first rank of Bethel and $^{c}$Anat-Bethel at Tyre in the seventh century B.C. has not been widely recognized. It is quite possible that Melqart, the city-god of Tyre, later superceded this ʾEl-deity and his consort, just as Marduk, city-god of Babylon, came to outrank the supreme god Anu in the Babylonian pantheon. But the position of Melqart as "lord of Tyre" (*b$^{c}$l ṣr*)--although the phrase does not unambiguously denote supremacy in the pantheon--does not contradict the evidence from the E/B god-list: this epithet is applied to Melqart no earlier than the second century B.C. (*KAI* 47).

Finally, one must bear in mind that the present study has been concerned with the gods of Carthage only with respect to their inclusion in the *official* pantheon. Caution should therefore be exercized in applying any of this evidence to the question of the popular religion of Carthage in 215 B.C.

## Appendix I:
## Charts

### Chart 1

*The Position of God-Lists*
*in First-Millennium Treaties*

| Treaty | After Introduction | At End of Text |
|---|---|---|
| M/Š | [ ] | x* |
| A/M | [ ] | x |
| Sf1 | x | |
| Senn | x | x |
| E/B | [ ] | x* |
| VTE | x x | x* |
| *ABL* 1105 | x | x* |
| A/Q | x | x |
| S/N | x | x* |
| H/P | x | |

*With curse(s) following

## Chart 2

*The "Canonical" God-List from Ur III*[1]

| | |
|---|---|
| $^{d}$INNANA | May Ištar |
| *an-nu-ni-tum* | "the Skirmisher," |
| AN | Anu, |
| $^{d}$*En-líl* | Enlil, |
| *Ìl-a-[b]*$a_4$ | Il-aba, |
| $^{d}$EN.ZU | Sin, |
| $^{d}$UTU | Šamaš, |
| $^{d}$*Nè-eri*$_{11}$*.gal* | Nergal, |
| $^{[d]}$*U-um* | Um ("Day"), |
| $^{[d]}$*Nin-kara* | Nin-karra(k), |
| *ì-lu* | the great |
| *ra-bí-ù-tum* | gods |
| *in* ŠU.NIGIN-*su-nu* | in their totality |
| *ar-ra-tám* | with a grievous |
| [*l*]*e-mu-ut-tám* | curse |
| [*l*]*i-ru-ru-uš* | curse him. |

## Chart 3

*The God-Lists in CH and M/Š*

| CH | M/Š |
|---|---|
| | Marduk* |
| | Nabu* |
| Anu* | [Anu]* |
| Enlil* | [Enlil]* |
| Ninlil* | Ninlil* |
| Ea* | Ea* |
| Šamaš* | Šamaš* |
| Sin* | [Sin]* |
| Adad* | [Adad]* |
| Zababa* | [Zababa]* |
| Ištar* | (text breaks off) |
| Nergal* | |
| Nintu* | |
| Nin-karra(k)* | |
| the great gods of heaven and the netherworld* | |
| Enlil* | |

*With curse(s) following

## Chart 4

*The God-List in A/M*

6 Assur, King of heaven and earth

7 Anum, Antum

Enlil, Ninlil

8 Ea, Damkina

Sin, Ningal

9 Šamaš, Aya

Adad, Šala

10 Marduk, Ṣarpanitum

Nabu, Tašmetum

11 Ninurta, Gula

Uraš, Belat ekalli

12 Zababa, Bau

Nergal, Laṣ

13 Madanu, Ningirsu

14 Ḫumḫummu, Išum

15 Erra, Nusku

Ištar, Lady of Nineveh

16 Ištar, Lady of Arbela

17 Adad of Kurba-il

18 Adad of Aleppo

19 Palil, "He who marches in front"

20 the Sebetti, the warriors

21 (the beginning of the listing of the gods of the vassal party)

## Chart 5

*The List of Divine Witnesses in Sf1*

| | |
|---|---|
| [*qdm* DN] *wmlš* | [Before DN] and *mlš*, |
| *wqdm mrdk wzrpnt* | before Marduk and Ṣarpanitum, |
| *wqdm nbʾ wt*[*šmt* | before Nabu and Ta[šmetum, |
| *wqdm ʾr wnš*]*k* | before Erra and Nus]ku, |
| *wqdm nrgl wlṣ* | before Nergal and Laṣ, |
| *wqdm šmš wnr* | before Šamaš(-)and(-)Nur, |
| *wqdm s*[*n wnkl* | before Si[n and Nikkal, |
| *wq*]*dm nkr wkdʾh* | be]fore Nikkar and *kdʾh*, |
| *wqdm kl ʾlhy rḥbh* *wʾdm* [. . . | before all the gods of RHBH and ʾDM |
| *wqdm hdd ḥ*]*lb* | [before Hadad of A]leppo, |
| *wqdm sbt* | before the Sebetti; |
| *wqdm ʾl wᶜlyn* | before ʾEl-and-ᶜElyon, |
| *wqdm šmy*[*n wʾrq* | before Heav[en and Earth, |
| *wqdm mṣ*]*lh wmᶜynn* | before Aby]ss and Springs, |
| *wqdm ywm wlylh* | before Day and Night. |

| | |
|---|---|
| *śhdyn kl ʾ*[*lhy ktk* *wʾlhy ʾrpd*] | All the g[ods of KTK and the gods of Arpad] are witnesses; |
| *pqḥw ᶜynykm lḥzyh ᶜdy* *br gʾyh* [*ᶜm mtᶜʾl* *mlk ʾrpd*] | Open your eyes to look upon the treaty of Bir-Gaʾyah [with Matiᶜel, the king of Arpad!] |

Chart 6

*Comparison of the Structures of the List of Divine Witnesses in Sf1 and H/P*

| Section | Sf1 | H/P |
|---|---|---|
| A. High gods | | |
| 1. supreme deities | [DN] and *mlš* | Zeus and Hera |
| 2. other high gods | Marduk . . . Sebetti | Apollo . . . Poseidon |
| B. Gods Paralleled in the Hittite treaty tradition | 'El-and-ʿElyon | Gods of mercenaries |
| | Three pairs of of natural phenomena | Two triads of natural phenomena |
| C. Concluding summary phrases | All the g[ods of KTK and Arpad] | All the gods of Carthage, All the gods of Macedonia, etc., All the gods of the army |
| D. Statement/ invocation of divine attestation | [All the gods are] witnesses Open your eyes to look upon the treaty . . . | . . . who are present (as witnesses) at (the conclusion of) this treaty |

## Chart 7

*The God-Lists in the Sennacherib Treaty-Fragment (VAT 11449)*[2]

### Obverse

2 [$^{d}$*Aš-šur* $^{d}$*Nin-líl* $^{d}$]XXX $^{d}$*Nin-gal* $^{d}$UTU [$^{d}$*A-a*]

3 [ ? ? ? $^{d}$*E*]*n-líl* $^{d}$IM $^{d}$*Ša-la* [$^{d}$*Kip-pat*.KUR]

4 [$^{d}$*Be-let* DINGIR.MEŠ?] $^{d}$[*Nin-ur*]*ta*? $^{d}$XV *aš-šur*[$^{ki}$-*i-tu*][3]

5 [$^{d}$*Za-ba*$_4$-*ba*$_4$ $^{d}$*Ba-ba*]$_6$? $^{d}$UGUR DINGIR.MEŠ [GAL.MEŠ]

6 [*ár-rat la nap-šur ma-ru-u*]*š-tu li*[-*ru-ru-ku-nu*]

2 [May Assur, Ninlil,] Sin, Ningal, Šamaš, [Aya,]

3 [ ? ? ? E]nlil, Adad, Šala, [Kippat-mati,][4]

4 [Belet-ilani(?),] Ninurta(?), the Assyr[ian] Ištar,

5 [Zababa, Ba]u(?), Nergal, the [great] gods,

6 c[urse you with an indissoluble, balef]ul [curse].

### Reverse

7 [$^{d}$*Aš-šur* $^{d}$*Nin-líl* $^{d}$]XXX $^{d}$*Nin-gal* $^{d}$UTU $^{d}$[*A-a* ? ? ]

8 [ ? $^{d}$*En-líl* $^{d}$IM $^{d}$*Š*]*a-la* $^{d}$*Kip-pat*.KUR [$^{d}$*Be-let* DINGIR.MEŠ?]

9 [$^{d}$XV *ša* $^{uru}$NINA$^{ki}$] $^{d}$XV *ša* $^{uru}$[LÍMMU.DINGIR$^{ki}$]

10 [$^{d}$*Za-ba*$_4$-*ba*$_4$ $^{d}$*Ba-b*]*a*$_6$ $^{d}$MAŠ $^{d}$EN⸢ŠADA⸣ [ . . . ? ? ]

11 [DINGIR.MEŠ GAL.MEŠ *ár-r*]*at la nap-šur ma-r*[*u-uš-tu*]

12 [*li-ru-ru-ku-nu*]

7 [May Assur, Ninlil,] Sin, Ningal, Šamaš, [Aya ? ? ]

8 [ ? Enlil, Adad, Š]ala, Kippat-mati, [Belet-ilani(?),]

9 [Ištar of Nineveh,] Ištar of [Arbela,]

10 [Zababa, Ba]u, Ninurta, Nu⸢sku,⸣ [ . . . ? ? ]

11 [the great gods, curse you with] an indissoluble

12 bal[eful cur]se.

## Chart 8

*The God-list in E/B*

(upper part of tablet broken off)

Column IV

| | | |
|---|---|---|
| 1 | [NINLIL(=Ištar) of Nineveh]* | = VTE 457-58 |
| 2 | Ištar [of Arbela]* | = VTE 459-60 |
| 3 | Gula* | = VTE 461-63 |
| 5 | the Sebetti* | = VTE 464-65 |
| 6 | Bethel, ᶜAnat-Bethel* | cf. VTE 466-68 |
| 8 | the great gods of heaven and the netherworld, | cf. VTE 472-75 |
| | the gods of Assyria, | |
| | the gods of Akkad, | |
| 9 | the gods of "Trans-Euphrates"* | |
| 10 | Baᶜal-Šamem, Baᶜal-Malage, Baᶜal-Ṣaphon* | |
| 14 | Melqart, 'Ešmun* | |
| 18 | ᶜAštart* | cf. VTE 453, 428-30 |

*With curse(s) following

## Chart 9

*Three God-Lists in VTE*

| (A) | (B) | (C) |
|---|---|---|
| 13 In the presence of | | |
| 14 Jupiter, Venus, | | |
| 15 Saturn, Mercury, | | |
| Mars, Sirius; | | |
| 16 In the presence of | | |
| Assur | 25 Assur | 414 Assur* |
| Anu | 26 Anu | |
| Enlil | Enlil | 417 Ninlil* |
| Ea | Ea | 418a Anu*[6] |
| 17 Sin | 27 Sin | 419 Sin* |
| Šamaš | Šamaš | 422 Šamaš* |
| Adad | Adad | 425 Ninurta* |
| | | 428 Venus* |
| | | 431 Jupiter* |
| Marduk | Marduk | 433 Marduk* |
| 18 Nabu | 28 Nabu | 435 Sarpanitum* |
| Nusku | Nusku | 437 Belet-ilani* |
| Uraš | Uraš | 440 Adad* |
| | | 453 Ištar* |
| Nergal | Nergal | 455 Nergal* |
| 19 Ninlil | 29 Ninlil | |
| Šerua | Šerua | |
| Belet-ilani | Belet-ilani | |

Chart 9

| (A) | (B) | (C) |
|---|---|---|
| 20 Ištar of Nineveh | 30 Ištar of Nineveh | 457 NINLIL of Nineveh* |
| Ištar of Arbela | Ištar of Arbela | 459 Ištar of Arbela* |
| | | 461 Gula* |
| | | 464 the Sebetti* |
| | | 466 FOREIGN GODS* |
| 21 the gods of heaven and the netherworld | 31 All the gods of Assyria | 472 the great gods of heaven and the netherworld*[7] |
| | 32 All the gods of Nineveh, | |
| | 33 All the gods of Calah, | |
| | 34 All the gods of Arbela, | |
| | 35 All the gods of Kakzi, | |
| | 36 All the gods of Harran, | |
| 22 the gods of Assyria, | 37 All the gods of Assyria; | |
| | 38 All the gods of Babylon, Borsippa, Nippur, | |
| Sumer, and Akkad | 39 All the gods of Sumer and Akkad; | |
| 23 all the gods of the lands | 40 All the gods of the lands; | |
| | 41 All the gods of heaven and the netherworld; | |
| | 41a All the [go]ds of his land(s and his districts.[5] | |

*With curse(s) following

## Chart 10

### *The God-Lists in* ABL *1105*

**Obverse**

1 [ ] $^{d}$*Be-le*[*t*-DINGIR.MEŠ? ]

2 [ ]-*nu* DINGIR.MEŠ GAL.MEŠ [*šá* AN$^{e}$ (*ù*) KI$^{tim}$]

1 [ ] Bele[t-ilani? ]

2 [ ], the great gods [of heaven and the netherworld]

**Reverse**

| | |
|---|---|
| 5 AN.ŠÁR* | Assur |
| 6 $^{d}$AMAR.UTU* | Marduk |
| 7 $^{d}$AG* | Nabu |
| 8 $^{d}$<UTU>*[8] | Šamaš |
| 11 $^{d}$XXX* | Sin |
| 13 $^{d}$*É-a** | Ea |
| 15 $^{d}$IM* | Adad |
| 17 $^{d}$*Nin-urta** | Ninurta |
| 18 $^{d}$UGUR* | Nergal |
| 19 $^{d}$*Za-ba*$_{4}$*-ba*$_{4}$* | Zababa |
| 20 $^{d}$IGI.DU* | Palil |
| 21 $^{d}$*Ṣar-pa-ni-tum** | Ṣarpanitum |
| 22 $^{d}$*Na-na-a** | Nana |
| 24 $^{d}$INNIN *a-ši-bat* | Ištar of Arbela |

LÍMMU.DINGIR$^{ki}$

(text breaks off)

*With curse(s) following

## Chart 11

*The God-Lists in A/Q*

Obverse

3′ [$^{d}$*Aš-šur*]    $^{d}$*Ni*[*n*]-*lîl*    $^{d}$*x*[    ][9]

3′ [Assur,]    Ni[n]lil,    [ ? ? ]

Reverse

8′ [$^{d}$*Aš-šur* $^{d}$*Ni*]*n-lîl* $^{d}$XXX $^{d}$UTU

9′ [$^{d}$EN $^{d}$P]A $^{d}$[XV] [*šá* $^{(uru)}$] NINA[$^{ki}$ $^{d}$X]V

10′ [šá LÍMMU.DINGIR$^{ki}$ $^{d}$] U+GUR *x*[    ]

8′ [Assur, Ni]nlil, Sin, Šamaš,

9′ [Bel ( = Marduk), Na]bu, [Ištar] [of] Nineveh, [Iš]tar

10′ [of Arbela,] Nergal, [ . . . ]

## Chart 12

_The Fragments of the S/N Treaty_ (BRM _4.50_)

### Obverse

1 [_a-de-_]_e_⌉ _ša_ $^{md}$XXX.MU.S[I.SÁ . . . ]
2 [DU]MU $^{m}$AN.ŠÁR.DÙ.A [MAN ŠÚ MAN KUR AN.ŠÁR]
3 [T]A* $^{md}$AG.[A]-_na_ [ ]
4 [T]A* $^{d}$?AB-[ ]A [ ]
5 [T]A* _Aq-ri-_[? . . . _iš-ku-nu-ni_[10] ]
6 [_i-n_]_a ma-ḫar_ $^{d}$S[AG.ME.GAR $^{d}$_Dil-bad_[11] . . . ]

1 [The treat]y which Sin-šum-l[išir,. . . ]
2 [s]on of Assurbanipal, [king of the world, king of Assyria,]
3 [wi]th Nabu-aplu-iddina [ ]
4 [wi]th ? ? ? ? [ ]
5 [wi]th Aqri[ . . . made ]
6 [i]n the presence of J[upiter, Venus . . .]

### Reverse

8 [ ] EN? AD?[ ]
9 [ ]_a-a_⌉_-bi ra-im a-bat ša_ [LUGAL-_u-ti-ja?_][12]
10 [ ]_x-ú_ GIM!? ⌈DUMU⌉?[13] _a-ra-an-š_[_u(-nu-ti?)_]
11 [_man-nu_] _ša ma!-mit ṭup-pi an-nê-e_ ⌈_e_[_-nu-u (e-gu-u?)_]
12 [_a-de-_]_e_ DINGIR.MEŠ GAL.MEŠ _i-pa-s_[_a-su_[14] DINGIR.MEŠ ]
13 [GAL.MEŠ] _šá_ AN$^{e}$ _u_ KI$^{tim}$ _ar-rat la na_[_p-šú-ri_ ]
14 [GIG-_t_]_u li-ru!-ru-šú-nu_[15] _e-liš i-na_ T[I.LA.MEŠ ]
15 [_li-sa-ḫu-šú-_]_nu šap-liš i-na_ KI$^{tim}$ ⌈_e_[_-ṭim-ma-šú-nu_]
16 [A.MEŠ] _lu-u-ṣa-am-me_[16] $^{d}$XXX NANNAR AN[$^{e}$ _u_ KI$^{tim}$ ]
17 [_saḫar-_]_šup-pu ki-ma na-aḫ-lap-ti lu-u-ḫal-_⌈_lip_⌉_-šú_[_-nu_]
18 _ma!-za-sa-šú!-nu_[17] TA* ŠÀ É.KUR É.GAL _lu-ḫal-liq_ [MU.MEŠ-_šú-nu_]
19 NUMUN.MEŠ-_šú-nu ṣa-lama-ni-šú-nu_[18] _i-na_ IZI _i-kar-ru-r_[_u_]
20 _di-pa-ra-šú-nu i-na_ A.MEŠ _ú-bal-lu_[_-ú_]

(Chart 12)

8 [ ] lord? [ ]

9 [ ] my enemy, (t)he(y) who love(s) [my royal?] word/command

10 [ ] like? a son? I will love t[hem/him? ]

11 [Who]ever ch[anges (or) disregards] the oath (written) on this tablet, (or)

12 eras[es the treat]y of the great gods, [may the great ]

13 [gods] of heaven and the netherworld with an indis[soluble, balefu]l curse

14 curse them; above among the l[iving]

15 [may they uproot th]em, below in the netherworld may they cause [their] g[hosts]

16 to thirst for [water.] May Sin, the luminary of heave[n and the nether-world,]

17 clot[he] th[em] with [lep]rosy like a cloak;

18 may he destroy their standing/service from temple and palace; [their progeny]

19 their posterity, (and) their images they will plac[e] in fire,

20 their torches they will extingui[sh] in water.

<u>Chart 13</u>

*Structure of the Typical*
*Neo-Assyrian Treaty God-List*

| | |
|---|---|
| Assur | supreme god of Assyrian pantheon; always named first |
| Anu | Assyrian high gods frequently named in royal/treaty god-lists; order of deities named is fairly consistent |
| Enlil | |
| Ea | |
| Sin | |
| Šamaš | |
| Adad | |
| Marduk | |
| Nabu | |
| Ištar of Nineveh | |
| Ištar of Arbela | |
| Gula | |
| the Sebetti | relatively infrequent; if named, always marks end of Assyrian gods; except when foreign gods are named, always followed immediately by summary phrase |
| (several foreign gods) | infrequently attested (VTE[C], E/B [and earlier A/M]) |
| the great gods (of heaven and the netherworld) | summary phrase; marks end of god-list as a whole (one exception: E/B where more foreign gods follow) |

## Chart 14

*Analyses of the Structure of the Hittite Treaty God-List*[19]

| Kestemont's Analysis | | Our Analysis |
|---|---|---|
| I. Essentially Masculine Deities | | I. Principal God(desse)s and Associated Deities |
| A. Principal Hittite Deities | sun-god, lord of heaven | A. Principal Hittite Deities |
| | sun-goddess of Arinna | |
| | storm-god of Hatti | |
| B. Dynastic Deities | Lulutašši | B. Dynastic Deities |
| C. Four Deities Connected with storm-god | Šeri, Ḫurri | C. Four Deities Connected with storm-god |
| | Nanni, Ḫazzi (mts.) | |
| D. Local storm-gods | storm-god of GN . . . | D. Local storm-gods |
| II. (Mostly) Feminine Deities | LAMMA | II. Protective Deities |
| A. Protective Deities | LAMMA of Hatti | |
| | LAMMA of Karaḫna | |
| | Zitḫariya | |
| | Karzi | |
| | Ḫapantaliya | |
| | LAMMA of the field | |
| | LAMMA of the shield | |

(Chart 14)

| | | |
|---|---|---|
| | | III. Telepinu and Related(?) Deities |
| | Lelwani, Ea | A. Related Deities |
| | Damkina | |
| B. Telepinu Cycle | Telepinu of Tawiniya | B. Telepinu Cycle |
| | Telepinu of Turmitta | |
| | Telepinu of Ḫanaḫana | |
| | Ašgašipa, Nisaba | C. Related(?) Goddesses |
| C. Five Sub-Groups | | IV. Oath-God(desse)s and Related Deities |
| 1. Sin | Moon-god, Lord of oath | A. Major Oath-Deities |
| 2. Išhara | Išḫara, Queen of oath | |
| 3. Ḫepat | Ḫepat, Lady of heaven | B. Hepat Cycle |
| | Ḫepat of Aleppo | |
| | Ḫepat of Uda | |
| | Ḫepat of Kizzuwanti | |
| 4. Ištar | | C. Ištar and Related Deities |
| | Ištar, Lady of the field | 1. Ištar Cycle |
| | Ištar, Lady of Nineveh | |
| | Ištar, Lady of Ḫattarina | |
| | Ninatta, Kulitta | 2. Related Deities |

(Chart 14)

| | | |
|---|---|---|
| 5. Some Adventitious Deities | (Taruppašani) | |
| | (Mammiš) | |
| D. Zababa Cycle | ZABABA | V. War Deities |
| | ZABABA of Hatti | |
| | ZABABA of Aziya | |
| | Iyarri | |
| | Zappana | |
| E1. (Very rare) | (Marduk) | |
| E2. Series of Local Deities | | VI. Local Deities |
| | Ḫantitaššu of Ḫurma | |
| | Apara of Šamuḫa | |
| | Kataḫḫa of Ankuwa | |
| | "Queen" of Katapa | |
| | (Am)mamma of Taḫurpa | |
| | Ḫallara of Dunna | |
| | Ḫuwaššanna of Ḫupišna | |
| | "Lady" of Landa | |
| | Kuniyawanni of Landa | |
| | (NIN.PISAN.PISAN.NA of Kinza) | |
| III. Gods of a General Character | | VII. Lowest-Ranking Gods Venerated in the Cult |
| A. Divinized Mountains | Mt. Lebanon | A. Divinized Mountains |

(Chart 14)

| | | |
|---|---|---|
| | Mt. Šariyana | |
| | Mt. Pišaiša | |
| B. Unattached Deities | Gods (of the?) Lulaḫḫi | B. Gods of Mercenaries |
| | Gods (of the?) Ḫapiri | |
| C. The Ensemble of Hatti-Gods | | VIII. Summary Statements |
| | All the gods (and) goddesses of Hatti | |
| | All the gods (and) goddesses of GN | |
| D. Cosmological Deities | Ereškigal | IX. "Non-Cultic" Witnesses |
| E. Traditional Deities | Nara, Napšara | A. The Olden Gods |
| | Minki, Amunki | |
| | Tušši, Ammizzadu | |
| | Alalu | |
| | Anu, Antum | |
| | Enlil, Ninlil | |
| | Belet-ekalli | |
| F. Divinized Physical Phenomena | Mountains, Rivers | B. Divinized Natural Phenomena |
| | Springs | |
| | the Great Sea | |
| | Heaven and Earth | |
| | Winds, Clouds | |
| | | X. Summons to Witness |

## Chart 15

*The Greek Text of the H/P God-List (Emended)*

Ἐναντίον Διὸς καὶ Ἥρας καὶ Ἀπόλλωνος,
ἐναντίον δαίμονος Καρχηδονίων καὶ Ἡρακλέους καὶ
Ἰολάου,
ἐναντίον Ἄρεως, Τρίτωνος, Ποσειδῶνος·

ἐναντίον θεῶν τῶν συστρατευομένων,
<ἐναντίον> Ἡλίου καὶ Σελήνης καὶ Γῆς,
ἐναντίον ποταμῶν καὶ λιμνῶν καὶ ὑδάτων·

ἐναντίον πάντων θεῶν ὅσοι κατέχουσι Καρχηδόνα,
ἐναντίον θεῶν πάντων ὅσοι Μακεδονίαν καὶ τὴν ἄλλην
Ἑλλάδα κατέχουσιν,
ἐναντίον θεῶν πάντων τῶν κατὰ στρατείαν,
ὅσοι τινὲς ἐφεστήκασιν ἐπὶ τοῦδε τοῦ ὅρκου.

## Chart 16

*Translation of the H/P God-List*

In the presence of Ba$^{c}$al-Ḥamon, Tanit, and Rešep,
In the presence of the protective deity of the
Carthaginians ($^{c}$Aštart), Melqart, and ’Ešmun,
In the presence of Ba$^{c}$al-Šamem (Hadad), Kušor, and
Ba$^{c}$al-Ṣaphon;

In the presence of the gods of those who take the
field with (us),
<In the presence of> Sun, Moon, and Earth,
In the presence of rivers, seas, and waters;

In the presence of all the gods who dwell in Carthage,
In the presence of all the gods who dwell in Macedonia
and the rest of Greece,
In the presence of all the gods of the army--

as many as are present (as witnesses) at (the
conclusion of) this treaty.

## Chart 17

*Table of Equivalences*

| Greek (H/P) | Carthaginian | Tyrian (E/B) | Ugaritic |
|---|---|---|---|
| Zeus | Ba$^c$al-Ḥamon | *Bayt-il* | 'El |
| Hēra | Tanit | *$^c$Anat-Bayt-il* | $^c$Anat |
| Apollōn | Rešep | ________ | Rešep |
| daimōn Karchēdoniōn | $^c$Aštart | *Aštart(u)* | $^c$Aṯtart |
| Hēraklēs | Melqart | *Milqart(u)* | Milk |
| Iolaos | 'Ešmun | *Ešumun(u)*[20] | ? |
| Arēs | Ba$^c$al-Šamem | *Baal-šamem(e)* | Haddu |
| Tritōn | Kušor | *Baal-malagê* | Kôṯar |
| Poseidōn | Ba$^c$al-Ṣaphon | *Baal-ṣapun(u)* | Ba$^c$lu-Ṣapāni (= Haddu) |

## Notes

1. See Roberts, *Semitic Pantheon* 145-64. The list here is from *PBS* v 36 Rev.IV.

2. For a god-list from Sennacherib's inscriptions similar to these two, see Luckenbill, *Sennacherib* 142.13-15.

3. *Pace* Ebeling(*Stiftungen,* 9)*aš-šur* cannot be the god Assur. See Chapter III n. 36.

4. On this goddess, see Frankena, *Tākultu* 97-98. She is mentioned in Luckenbill, *Sennacherib*, where the name is misread as $^d$GAM.LAT.

5. This line is found only in fragments 45A and 45E; see Borger, "Asarhaddon-Verträgen" 176.

6. This entry (with epithets and curse) appears only in fragments 29, 35, and 50S; see Borger, ibid. 187.

7. Another series of deities follows this third god-list in VTE. On this see Chapter III.

8. The UTU sign is missing here, probably through haplography; read  etc. for .

9. Reading uncertain; perhaps Sin (note the sequence Assur, Ninlil, Sin, etc. on the reverse).

10. Judging from VTE 12, *iškunū-ni* should go here, not in line 6 as Borger suggests ("Aufstieg" 76).

11. According to Scheil's edition ("Sin-šar-iškun" 47) the sign following the divine determinative seems to be UTU (); the text breaks off after this point. Clay's copy has only the upper part of a vertical wedge after the DINGIR sign. But since VTE's format is identical up to this point to that of S/N, we might read 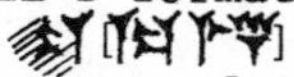; thus [*i-n*]*a ma-ḫar* $^{d}$S[AG.ME.GAR $^{d}$*Dil-bad* etc.], "[i]n the presence of J[upiter, Venus, etc.]," as in VTE. Cf. the "oath of fealty to Assurbanipal" (*ABL* 1239), where [. . .] $^{d}$SAG.ME.GAR $^{d}$*Dil-bad* [. . .] are named in what may well have been part of a list of divine witnesses.

12. Restoration conjectural.

13. Restoration conjectural.

14. Lines 11-12 have been restored on the basis of VTE 397-98.

15. Clay's copy reads *li-ri-im-šú-nu*; Scheil's, *li-ri-ru-šú-nu*.

16. Lines 12b-16a have been restored on the basis of VTE 472, 275-77; cf. *ABL* 1105 Rev 14 and CH Rev xxvii 34-40.

17. For this reading see J. Nougayrol, "*Sirrimu* (non **purîmu*) 'âne sauvage'," *JCS* 2 (1948) 207 n. 15.

18. For this reading see *AHW* 1078b.

19. The list of Hittite gods here is representative rather than exhaustive. It is derived from a number of treaty god-lists. For a complete list of the deities found in these god-lists, see Kestemont, "Panthéon" 156-69.

20. The name is transliterated by Borger as $^{d}$*Ia-su-mu-nu* (*Inschriften* 109 and pl. 4). However, the name is probably to be read **Ešumun*. As regards the first syllable, the cuneiform signs *e* and *ia* are quite similar: *e* = , *ia* = 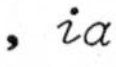. (Note the similar misreading of the divine name $^{d}$*E-ri-qi* as $^{d}$*Ia-ri-qi* [Frankena, *Tākultu* 87].) Thus we would read $^{d}$*E!-su-mu-nu*, i.e., **Ešumun(u)*, which would better correspond to the Phoenician form *ʾšmn* as well as non-Phoenician transcriptions of the name (cf. Greek *Esmounos*, *Esym-*, *-ysmoun*; Latin *Asmunis*, *-ismunis*, *-usmyn* [see Benz, *Personal Names* 279]). As regards the sibilant, it is now recognized that in the Assyrian dialect the grapheme *su* was pronounced /*šu*/, *sa* was pronounced /*ša*/, etc. (see S. A. Kaufman, *The Akkadian Influences on Aramaic* [Assyriological Studies 19; London/Chicago: University of Chicago Press, 1947] 140 and S. Parpola, "The Alleged Middle/Neo-Assyrian Irregular Verb **naṣṣ* and

the Assyrian Sound Change *š* > *s*," *Assur* 1/1 [1974] 1-10). In this same god-list Ba^c^al-Šamem is written ^d^*Ba-al-sa-me-me* and ^c^Aštart, ^d^*As-tar-tú*; note also ^c^Atarsamain written ^d^*A-tar-sa-ma-a-a-in*. The final vowels in the divine names here (^d^*Ba-al-sa-me-me*, ^d^*Ba-al-sa-pu-nu*, ^d^*Mi-il-qar-tu*, *^d^*E!-su-mu-nu*, ^d^*As-tar-tú*) were probably imitative of Akkadian case-endings and not pronounced.

Appendix II

## The Question of Foreign Gods in Mesopotamian Treaties

The question of whether foreign deities are involved as witnesses or guarantors of the oath in Mesopotamian treaties is to some extent controverted. (By "foreign deities" here we mean gods of the vassal party.) Only one Assyrian treaty (E/B) preserves almost intact a list of what are without any doubt foreign gods--in this case, gods of the Tyrian pantheon.[1] But even here there is disagreement as to where the list of foreign gods begins.[2] How common was the listing of the vassal party's gods (or at least several of them) in the formulation of Mesopotamian treaties? The question is difficult to answer with any degree of certainty, because the endings of the majority of these lists are broken off or in an extremely fragmentary state. We shall here assess the evidence for the listing of foreign gods in the Mesopotamian treaties and treaty fragments known to date.

For the purpose of clarity we may divide the material to be discussed into several categories: (1) lists in which foreign gods are not represented and in which the drafters of the treaty did not include such gods; (2) lists broken off so that it is no longer possible to determine whether or not they ever mentioned foreign gods; (3) lists which may contain foreign gods; and (4) lists which clearly do mention such gods.

Two treaty texts belong to the first category. Senn may be either a treaty or a loyalty oath (Appendix I, Chart 7). In the two god-lists preserved all the gods mentioned belong to the Assyrian pantheon. In line 5 of the obverse the last preserved signs are DINGIR.MEŠ, most likely to be restored DINGIR.MEŠ [GAL.MEŠ], "the [great] gods," a summary phrase that commonly concludes Mesopotamian god-lists. The next line is to be restored: [*ár-rat la nap-šur ma-ru-u*]*š-tu li*[*-ru-ru-ku-nu*], the curse that most frequently accompanies the great gods. Hence this list is concluded without naming any foreign gods. The same is true of the list on the reverse, which likewise ends [DINGIR.MEŠ GAL.MEŠ *ár-r*]*at la nap-šur mar-r*[*u -uš-tu li-ru-ru-ku-nu*]. If this oath was administered to Assyrians (as in the case of *ABL* 1239) there would be no reason to invoke foreign gods.

Like Senn, S/N is in too fragmentary a state to conclude with certainty whether it is part of a treaty with a non-Assyrian nation (Appendix I, Chart 12). Judging from the names preserved on the obverse (3-5)--Nabu-apla-idinna and Aqri-[. . .]--the parties with whom the pact was concluded seem to be Assyrian or Babylonian. There is a good chance, then, that this is a loyalty oath and not an international treaty. As we have noted, S/N has on the obverse (which preserves the first lines of the document) a text quite similar to the beginning of VTE. Like VTE 13 it appears to contain the beginning of a god-list, which unfortunately is lost except for a few traces. Given the formal correspondence of S/N with VTE up to this point, one would hardly expect to find foreign deities in S/N; note that neither VTE(A) nor VTE(B) contains such deities. On the reverse of S/N is a section very similar to VTE (compare S/N Rev 11-12 with VTE 397-98), warning against changing or ignoring the treaty-oath. In S/N there follow several curses invoking "[the great gods] of heaven and the netherworld" and Sin. This section is preserved almost intact and names no foreign gods.

The fragmentary condition of the god-list(s) in three other treaty texts makes it impossible to determine whether these ever contained the names of foreign gods. What remains of the list of gods and accompanying curses in M/Š is almost identical to those in the epilogue of CH (see Appendix I, Chart 3). Since Babylon appears to be the dominant party here,[3] any specifically Assyrian deities that may have been invoked on this occasion would most likely be found at the end of the list. But the text is broken before the listing of the official gods of Babylon is completed.

*ABL* 1105 is unique insofar as the oath is cast in the first person plural (Appendix I, Chart 10). Most likely it is specifically a loyalty oath[4] rather than a treaty with a foreign power. The fact that the script is Babylonian rather than Assyrian, the repeated mention of Šamaš-šum-ukin (Obv 6, 29), and the fact that Marduk and Nabu are listed immediately after

Assur in the concluding god-list (Rev 6-7) all suggest that this is a loyalty oath sworn by Babylonians (probably to Assurbanipal after his succession to the throne). Since the official pantheon of Assyria was not distinct from that of Babylon (with the exception of the god Assur and the Sebetti, who were not "great gods" in Babylon), there can be no question of "foreign gods" in this text.

A/Q contains two god-lists (Appendix I, Chart 11). The first is quite brief and almost certainly contains nothing but Assyrian gods. The concluding list is longer but broken off at the end; the names preserved are all of Assyrian gods. Thus we cannot tell if the text contained the names of one or more of the deities of Qedar. It is interesting that Deller and Parpola ("Vertrag Assurbanipals" 464) restore Obv 2 as: [DINGIR.MEŠ KUR] *Aš-šur* KUR *Qi-id*[*-ri* . . .], "[the gods of the land of] Assyria (and) the land of Qed[ar . . .]." But this restoration is at best conjectural and provides no evidence that the gods of Qedar are mentioned in this treaty.

Of the five texts surveyed thus far only two (M/Š and A/Q) are clearly treaties with a foreign power; and in both of these the concluding god-list, which is where foreign deities are listed if they are listed at all,[5] is broken before its conclusion.

We shall now consider treaties between Assyria and a foreign power which may contain foreign gods. A/M lists thirty-seven deities of the official Assyrian pantheon (counting the Sebetti here as one deity for the sake of enumeration) as oath-gods (Appendix I, Chart 4). After the Sebetti at least five more deities are listed (lines 21ff), although the text has suffered damage at this point. These are almost certainly the gods of the vassal party (Mati'ilu of Bīt-Agusi).[6] The first deity appears to be Dagan,[7] who would thus be the chief god of Mati'ilu's official pantheon. Moreover the endings of the geographical names preserved in this section do not appear to be those of Assyrian cities.[8] Finally, if these are not foreign gods, one must explain why another Ištar-goddess ($^{d}$XV) was added in line 23 after two such goddesses had already been invoked in lines 15-17; and why another storm-god ($^{d}$IM) was added in line 24 after three such gods had been listed in lines 9, 17-18.

There is somewhat less agreement on the question of foreign gods in VTE(C). After mentioning the Sebetti (464-65) and before the concluding summary phrase, "the great gods of heaven and the netherworld" (472-75), the final god-list in VTE contains one or more entires of uncertain length[9] which are in a poor state of preservation. Frankena believes that deities of the vassal party are listed here.[10] Recently, however, M. Cogan has argued that foreign gods were invoked only in those Assyrian treaties not concluded with provinces of Assyria, although his understanding of the contents of VTE 466-71 is not clear.[11] But even if Cogan's thesis is correct, the lines in question may still name foreign gods. The locales of the vassal parties do not indicate that these vassal-treaties of Esarhaddon were concluded exclusively with provinces. Of the seven place-names preserved,[12] only three were clearly Assyrian provinces: Zamua, Elpa, and Sikrisu.[13] Urakazabana, mentioned among other tribute-bearing areas in Media,[14] may also have been a province. The remaining three areas--Naḫšimarti, Karzitali, and Izai--are unlocated.

No conclusion can be reached on the foreign god question until the lines in question are analyzed. We therefore reproduce these here, including the lines immediately preceding and following:

a. Fragment X17:
464 $^{d}$*Se-bet-ti* DINGIR.ME[Š *qa*]*r-*[*du-te ina* $^{giš}$TUKUL.MEŠ-*šú-nu*]
465 *ez-zu-ti na-aš-pan*[*-ta-ku-nu liš-kun*][15]
466 [$^{d}$*Ga*]*r?-ga-miš ri*(sic)*-im-ṭ*[*u dan-nu* etc.][16]

464 May the Sebetti, [the war]r[iors, with their] fierce
465 [weapons bring about your] downf[all.]
466 [May (the god of?) Ca]rchemish [put a strong] *rimṭ*[*u* etc.][17]

b. Fragment 37:
467 $^{d}$[. . .][18]
468 *ina* ŠU$^{II}$ [UR.]M[AḪ *a-ki-li li-mal-li/lu-ku-nu*][19]
469 $^{d}$XV ⸢$^{uru?}$⸣ Ti[-[20] . . . $^{d}$? $^{uru}$]*Gar-ga-miš*
470 *ri-im-ṭu dan-nu ina* [ŠÀ]-*ku-nu* ⸢*liš*⸣-*kun* [ÚŠ].MEŠ-*ku-nu*
471 *ki-ma ti!-ki*[21] *ana qaq-qar lit-ta-rad*
472 DINGIR.[MEŠ G]AL.MEŠ-*t*[*i*] *šá* AN$^{e}$ KI$^{tim}$ *a-ši-bu-tu kib-ra-*[*a-ti*]
473 *ma-la ina ṭup-pi an-né-e* MU-*šú-nu zak-ru*
474 *lim-ḫa-ṣu-ku-nu li-kil-mu-ku-nu*
475 *ár-ra-tu ma-ru-uš-tu ag-giš li-ru-ru-ku-nu*

467 [May] (the god) [X . . .]
468 [Hand you over] to the power of a [devouring] li[on].
469 May IŠTAR[22] of [the city] Ti[? . . . (and) the god X of] Carchemish

470 place a strong *rimṭu* inside of you; may your [blood]
471 flow down to the ground like a torrent of rain.
472 May the [gr]eat god[s] of heaven (and) the netherworld, who dwell in the world,
473 as many as are named on this tablet,
474 strike you down, look with disfavor upon you,
475 (and) curse you angrily with a baleful curse.

Several considerations support the view that foreign deities are in fact listed in lines 466-71. The first is structural. These entries come between the Sebetti and the summary phrase. As we have already seen in the case of Sf1 and A/M, the Sebetti mark a caesura within the list. Furthermore, although the Sebetti are not included in the majority of official god-lists from Assyria, it is significant that whenever they are named in such a list, at least from the time of Sargon II (721-705) on,[23] they always appear last among the Assyrian gods; the only entry to appear after them is a summary phrase, usually "the great gods of heaven and the netherworld" or something similar. For example, the Cyprus Stela of Sargon II begins with an official list of Assyrian gods, ending with "the Sebetti [. . .] (and) the great gods who manage heaven and the netherworld."[24] The Bavian Inscription of his successor, Sennacherib (704-681), also begins with a god-list that ends "the Sebetti, (and) the great gods."[25] A number of official god-lists in the inscriptions of Esarhaddon (680-669) also show these two entries last and in juxtaposition: (1) "the Sebetti, the divine warriors . . . (and) the great gods who dwell in heaven and the netherworld";[26] (2) the Zinčirli Stela: "the Sebetti, the divine warriors . . . (and) all the great gods";[27] (3) a stela from the Nahr-el-Kalb: "[the Sebetti, (and)] all the [great god]s";[28] (4) a Neo-Assyrian text ascribed by Borger to Esarhaddon: "the Sebetti, the divine warriors . . . (and) the great gods, lord[s of heaven and the netherworld]."[29]

The function of these two entries in the god-lists is clear: the Sebetti conclude the list of Assyrian gods and the summary phrase concludes the list as a whole. In lists containing only Assyrian deities, therefore, nothing intervenes between these two. Only in the treaty texts--A/M, VTE(C), and E/B--are more gods listed immediately after the Sebetti. In the case of A/M, as we have seen, these additional deities are almost certainly those of the vassal party. Several other Mesopotamian documents also show that foreign gods were traditionally listed in this slot--that is, after the state gods of the superior party. A letter from Esarhaddon to the Elamite king Urtaku (*ABL* 918) contains a short god-list (9-11). The list is identical to those in a number of Esarhaddon's inscriptions,[30] with the exception of the last entry:

> Assur, Sin, Šamaš, Bel (Marduk), Nabu, Ištar of Nineveh, Ištar of Arbela, Manzinir.

Manzinir was one of the high gods of Elam.[31] Note how this foreign deity is named after the Assyrian gods.

The same pattern is attested in royal documents from Babylonia. A *kudurru* from the time of Nebuchadnezzar I (1126-1105) contains curses invoking not only several of the official gods of Babylon but also a number of deities of the party sworn to uphold the terms of the royal grant described on the boundary stone.[32] In the document Nebuchadnezzar frees Ritti-Marduk and the towns belonging to him from all jurisdiction of the rulers of the land of Namar. The imprecations designed to prevent violation of the grant are therefore directed at Namar as well as Bīt-Ḫabban, a Kassite tribal sub-group[33] in or around Namar. Because of its importance for our study we quote the god-list in its entirety:

37 May *the great gods*, as many as
are named in heaven and the netherworld,
38 angrily curse that man.
May *god and king* look upon him with anger.
39 May *Ninurta*, the king of heaven and the netherworld,
and *Gula*, the bride of Ešarra,
40 destroy his boundary-stone and obliterate his seed.
41 May *Adad*, the canal-inspector of heaven and
the netherworld,
the lord of springs and rain,
42 fill his canals with mud . . . [more curses]
46 May *Šumalia*, the lady of the bright mountains,
47 who dwells upon the summits,
who treads beside the springs,
48 $^{d}$*IM*, $^{d}$*UGUR*, and *Nana*,
*the gods of Namar*--
49 May *Šaḫan*,[34] the bright god,
the son of the temple of Der,

50 $^{d}$*XXX*, and *Belet-Akkadi*,

*the gods of Bīt-Ḫabban*--

51 May *these great gods* in the anger of their hearts

52 ever think of him with evil intent.

Note that the entries "the great gods," "god and king," the gods Ninurta and Gula, and Adad are each associated with one or more curses. But the gods immediately following, beginning with Šumalia, are all associated with only one curse; this is a clear indication that these gods are connected in some way. As a matter of fact they are the deities of Namar and Bīt-Ḫabban, and are explicitly designated as such. Although Šumalia and her spouse Šuqamuna are sometimes designated "the gods of the king"[35] and are frequently mentioned on Kassite boundary-stones, she is not mentioned on any other post-Kassite *kudurru*. This, plus the fact that she is unmistakably connected with the following deities by the curse (51-52), means that Šumalia, usually considered to be a Kassite goddess,[36] is here ranked among the gods of Namar. The gods whose names are written $^{d}$IM and $^{d}$UGUR are most likely not Adad (already named) and Nergal but their Kassite equivalents.[37] The "gods of Bīt-Ḫabban" include not only $^{d}$XXX (probably the Kassite moon-god, not the Babylonian Sin) and "the Lady of Akkad"[38] but also Šahan.[39] Thus lines 46-52 contain two parallel sub-sections of *foreign gods:* each names a deity with epithets followed by several gods without epithets, concluding with "the gods of GN." The overall structure of this god-list is clear: a series of gods of the superior party (Babylon), followed by a series of gods of the inferior party (Namar/Bīt-Ḫabban), followed by a summary phrase ("these great gods")--the same structure as in VTE 414-75.

A second clue to the nature of the deities in VTE 466-71 is the name *Gàr-ga-miš* in line 469; it is also possible, though by no means certain, that this city-name is also to be read in line 466. Judging from the structure of the other entries in this god-list, there is no escaping the fact that "Carchemish" is part of a divine epithet. Thus VTE 469-71 is an invocation of a deity of Carchemish. Since Carchemish was a province by the time this treaty was concluded,[40] Cogan's theory that the gods of provincial areas were not included in Assyrian treaty god-lists cannot be accepted.

One Assyrian treaty that has come down to us clearly lists foreign gods--E/B (Appendix I, Chart 8). The end of the document has fortunately suffered only minor damage. Lines 10-19 name six Tyrian gods. The controversy concerns lines 6-7, which name "Bethel (and) ᶜAnat-Bethel." Are these to be considered foreign gods? At first glance this would not seem to be the case. The summary phrase invoking "the great gods of heaven and the netherworld, the gods of Assyria, the gods of Akkad, (and) the gods of Trans-Euphrates" appears to mark a caesura in the list between the Assyrian and non-Assyrian gods. Hence one can understand why Teixidor insists that the E/B god-list "clearly divides into two sections" and that "the whole structure of the treaty shows clearly that in the Assyrian chancery Bethel and Anat-Bethel were not regarded as Phoenician gods."[41] Since, however, Teixidor makes no attempt to show why his analysis is "clearly" correct, one may safely conclude that his understanding of the structure of this god-list is based on little more than surmise. But if one compares this treaty god-list with other such lists, a different understanding emerges. We have already seen that in A/M, Sf1, and VTE(C) it is the Sebetti, not the summary phrase, who mark the dividing line. Moreover, it is clear from a comparison with god-lists in Neo-Assyrian inscriptions that the summary phrase always ends the list as a whole. Only in E/B are more deities mentioned immediately after the summary phrase. This means that there is something irregular about the E/B list; it should have concluded, like all other Neo-Assyrian lists, with the summary phrase. But for some reason more foreign gods are named. Was this done as an afterthought? Or in order to fill out the rest of the column? Or out of special deference to Tyre? In any case, despite the position of the summary phrase, a comparative analysis of the structure of this god-list shows that Bethel and ᶜAnat-Bethel occur precisely where we would expect the foreign gods to be named, and precisely where they are named in VTE(C). In fact, the curse accompanying these two gods in E/B is the same as that connected with the first foreign god in VTE 468.[42]

Aside from structural considerations there is other evidence that Bethel and ᶜAnat-Bethel are considered foreign gods here. The most important factor is that neither of them is ever found in any Neo-Assyrian god-list--royal, official, cultic, or otherwise. This fact at least suggests that the choice of these gods had something to do with the particular vassal party involved in this treaty. Secondly, all the Assyrian gods in the list bear one or more epithets. This is clear even in the case of the fragmentary lines in Rev IV.1-4, since these can be restored with certainty on the basis of VTE(C).[43] And since the missing portion of the Assyrian god-list in E/B was evidently identical with VTE(C) 414-65--to judge from the verbatim correspondence of the Assyrian gods, epithets, and curses in the extant portion of the E/B list--we may safely assume that the Assyrian gods listed there also bore epithets. But the foreign gods have no epithets. This is true not only of the gods listed after the summary phrase in E/B but of Bethel and ᶜAnat-Bethel as well. The absence of epithets connects these two deities with the other foreign gods. Thirdly, the list of Assyrian gods in VTE(C), E/B, and *ABL* 1105 names the deities one at a time--never severally. By contrast in VTE 469 several foreign deities appear to be mentioned together: "IŠTAR of (the city) G[N and DN of (the city)] Carchemish." And in E/B, Baᶜal-Šamem, Baᶜal-Malage, and Baᶜal-Ṣaphon are named together as are Melqart and 'Ešmun and, of course, Bethel and ᶜAnat-Bethel.

The weight of evidence, therefore, strongly supports the contention that Bethel and ᶜAnat-Bethel are the first-named foreign gods in the E/B god-list. Since the "foreign" or vassal party here is Tyre, these must be Tyrian gods. And since, as we have seen, official Neo-Assyrian lists of foreign gods invariably name the supreme deities in first place, Bethel and ᶜAnat-Bethel would be the supreme gods of the Tyrian pantheon at the time this treaty was concluded.

## Notes

1. See pp. 43-44.

2. See pp. 45-46.

3. Weidner, "Staatsvertrag" 27.

4. So apparently Wiseman, *Vassal-Treaties* 9.

5. See the discussion of A/M, VTE, and E/B immediately following.

6. So Weidner, "Staatsvertrag" 23; Cogan, *Imperialism and Religion* 47.

7. See p. 39.

8. Rev 21: [$^{uru}$[*m*]*u?-sur?-*⌈*u?*[*-n*]*a*
22: [ [*š*]*á?-ṁu-na*
23: [ ]*-ḫu-ḫa*

9. Borger, "Asarhaddon-Verträgen" 190.

10. Frankena, "Vassal-Treaties" 130.

11. *Imperialism and Religion* 47.

12. Wiseman, *Vassal-Treaties* 82.

13. Cogan, *Imperialism and Religion* 47 n. 29.

14. Ibid.

15. Restored from E/B IV.5; see Borger, "Asarhaddon-Verträgen" 190.

16. This is Borger's tentative restoration of this line (ibid.). Wiseman transliterates: [$^{d}$ . . . *ra-šid* EN IM . . .]. Where Borger restores [$^{d}$*Ga*]*r*(?)-*ga-miš*, there may be room for one sign after the DINGIR sign, which would be the name of the god here.

17. For the rest of this restored curse and a translation, see line 470 below.

18. If fragments X17 and 37 form a "join"--which Borger considers to be at least a possibility--this line would be the same as line 466 in fragment X17.

19. Restored from E/B IV.7; see Borger, ibid.

20. The reading of ⌈$^{uru}$⌉ here is not certain, but is at least possible judging from the copy. The sign following this, which would thus be the first syllable of the city-name, could also be read *An-d*[*i*] or *An-k*[*i*].

21. Following *AHW* 1357b and *CAD* "A"/2 217b.

22. The ideogram does not necessarily stand for the Assyro-Babylonian Ištar but can denote any goddess whom the Assyrians considered an Ištar-manifestation. Note similarly how the Arabian goddess $^{c}$Ataršamain could be designated by $^{d}$XV in an Assyrian inscription, K3405 Rev 6 (see Cogan, *Imperialism and Religion* 19-20).

23. The Sebetti also appear last in a listing of the official pantheon and just before a summary phrase in a royal grant issued by Adad-Nirari III (810-783); the list ends with "the Sebetti, (and) all these great gods of Assyria" (see Postgate, *Neo-Assyrian Royal Grants* 19).

24. See L. Messerschmidt and A. Ungnad, *Vorderasiatische Schriftdenkmäler der Königlichen Museen zu Berlin* 1 (Leipzig: 1907) no. 71.22-24 (pp. 65-66).

25. See Luckenbill, *The Annals of Sennacherib* 78.1.

26. See Borger, *Inschriften* §53.12-13.

27. Ibid. §65.10-11.

28. Ibid. §67.2 The restoration is all but certain given the verbatim correspondence of the extant portion of this god-list to the one in §65, except that in §67 the gods have no epithets.

29. Ibid., §85.I.20-21; see S. Langdon, *Tammuz and Ishtar: A Monograph upon Babylonian Religion and Theology* (Oxford: Clarendon Press, 1914) 186 and pl. 6.

30. Except for the last entry the list is identical more specifically to those in a number of passages from Esarhaddon's inscriptions; see Borger, *Inschriften* §§21.12; 27.1A.5-6; 1B.5-7; 2A.45.59; etc.

31. See Streck, *Assurbanipal* 1.clxviii n. 1.

32. *BBS* no. 6, col. iii.37-51.

33. See J. A. Brinkman, *A Political History of Post-Kassite Babylonia: 1158-722 B.C.* (AnOr 43; Rome: Pontifical Biblical Institute, 1968) 232.

34. On the reading of $^{d}$MUŠ as *Šaḫan*, see H. Wohl, "Niraḫ or Šaḫan?" *JANESCU* 5 (1973) 443-44.

35. See W. J. Hinke *A New Boundary Stone of Nebuchadnezzar I. from Nippur* (The Babylonian Expedition of the University of Pennsylvania; Series D: Researches and Treaties 4; Philadelphia: University of Pennsylvania Press, 1907) 57 (hereafter *NebNip*). Šuqamuna and Šumalia were therefore the gods of the Kassite royal family. See Brinkman, *Political History* 256.

39. That Šaḫan was an important deity to the "sons of Ḫabban" is evident from a land-grant from the district of Bīt-Ḫabban in which a "son of Ḫabban" gives his daughter a certain piece of property as her dowry; the bridegroom is made to swear "by the great gods *and the god Šaḫan*" not to raise a claim against the land in question. See *NebNip* 32.

40. Carchemish was made into an Assyrian province after a revolt in 718 B.C. See, for example, Saggs, *The Greatness That Was Babylon* 126.

41. Teixidor, *The Pagan God* 30-31.

42. See Borger, "Asarhaddon-Verträgen" 190.

43. See Borger, ibid. 189-90; Frankena, "Vassal-Treaties" 130.

## Notes

### Chapter I

1. F. W. Walbank, *A Historical Commentary on Polybius* (2 vols.; Oxford: Oxford University Press, 1967) 2.42.

2. See Chapter IV n. 504.

3. Livy, *Ab Urbe Condita Libri* 23.33-34.

4. "Hannibal's Covenant," *AJP* 73 (1952) 2.

5. Walbank, *Polybius* (Berkeley/Los Angeles: University of California Press, 1972), pp. 82-83; Bickerman, "An Oath of Hannibal," *TAPA* 75 (1944) 90.

6. Ibid. esp. 89, 94-95.

7. J. M. Moore, *The Manuscript Tradition of Polybius* (Cambridge: Cambridge University Press, 1965) 19.

8. Bickerman, "Hannibal's Covenant" 2.

9. See esp. Bickerman "Oath of Hannibal" 90-102.

10. Ibid. 96-101; idem, "Hannibal's Covenant" 6-8.

11. On this term, see pp. 8-9.

12. "Hannibal's Covenant" 10.

13. "Oath of Hannibal" 87.

14. Note, for example, Bickerman's remarks to the noted orientalist M. Weinfeld in a postcard: "Long time ago I wrote a paper: 'Asiatic origins of Greek treaty formulas,' but lacked the time to revise it. I am glad to see that you have tackled the same problem with knowledge of Oriental sources (Weinfeld, "$l^{c}$*nyyn mwnḥy bryt bywnyt wbrwmyt*" ["Greek and

Roman Covenantal Terms and Their Affinities to the East"], *Leš* 38 [1974] 237 n. 34).

15. See esp. "Hannibal's Covenant" 9-10.

16. Ibid.

17. This statement is based on an examination of the comprehensive collection of Greek treaties published by H. Bengtson (*Die Staatsverträge des Altertums. Zweiter Band: Die Verträge der griechisch-römischen Welt von 700 bis 338 v. Chr.* [Munich: C. H. Beck, 1962]) and H. H. Schmitt (*Die Staatsverträge des Altertums. Dritter Band: Die Verträge der griechisch-römischen Welt von 338 bis 200 v. Chr.* [Munich: C. H. Beck, 1969]). Hereafter these works will be referred to as *StVA* 2 and *StVA* 3 respectively.

18. "Hannibal's Covenant" 8.

19. Ibid. 6 n. 10.

20. *Historical Commentary* 44.

21. Ibid.

22. D. J. Wiseman, *The Vassal-Treaties of Esarhaddon* (London: British School of Archaeology in Iraq, 1958 [ = *Iraq* 20 (1958) Part I]) lines 1-13.

23. For example, the treaty between Sin-šum-lišir of Assyria and Nabu-aplu-iddina et al., ca. 625 B.C. (see Chapter III n. 14) and the treaty between Bir-Ga'yah of KTK and Mati$^{c}$el of Arpad, ca. 750 B.C. (see Chapter III n. 5).

24. Translation from W. R. Paton, *Polybius: The Histories* (LCL; Cambridge: Harvard University Press, 1923) 421.

25. In one of the most important monographs on *kudurrus*, F. X. Steinmetzer refers to them as "royal legal documents" (*Die babylonischen Kudurru* [*Grenzsteine*] *als Urkundenform* [Studien zur Geschichte und Kultur des Altertums 11/4-5; Paderborn: F. Schöningh, 1922] 96).

26. See D. J. McCarthy, *Treaty and Covenant: A Study in Form and in the Ancient Oriental Documents and in the Old Testament* (AnBib 21; Rome: Pontifical Biblical Institute, 1963) 104; revised edition (AnBib 21A; Rome: Pontifical Biblical Institute, 1978) 109; D. R. Hillers, *Treaty-Curses and the Old Testament Prophets* (BibOr 16; Rome: Pontifical Biblical Institute, 1964) 15.

27. "The Historical Development of the Mesopotamian Pantheon: A Study in Sophisticated Polytheism," *Unity and Diversity: Essays in the History, Literature, and Religion of the Ancient Near East* (ed. H. Goedicke and J.J.M. Roberts; Baltimore: Johns Hopkins University Press, 1975) 191.

28. See G. Komoroczy, "Das Pantheon im Kult, in den Götterlisten und in der Mythologie" *Or* 45 (1976) 81.

29. Ibid.

30. N. Oettinger, *Die Militärischen Eide der Hethiter* (StBoT 22; Weisbaden: Harrassowitz, 1976) 74.

31. O. R. Gurney, *Some Aspects of Hittite Religion* (Oxford: Oxford University Press, 1977) 6.

32. "Paralipomena Punica," *Cahiers de Byrsa* 6 (1956) 16.

33. On the non-correspondence of onomasticon and pantheon in the West Semitic world, see S. Moscati, *I Fenici e Cartagine* (Turin: Unione tipografico-editrice torinese, 1972) 533.

34. *Webster's New Collegiate Dictionary* (Springfield, MA: G. & C. Merriam, 1977) 828.

35. H. W. F. Saggs, *The Greatness That Was Babylon: A Sketch of the Ancient Civilization of the Tigris-Euphrates Valley* (New York/Toronto: New American Library, 1962) 321.

36. Ibid. 315-17.

37. Ibid.

38. On the "olden gods," see pp. 27-28.

39. See the treaty between Assur-nirari V of Assyria and Mati'ilu of Arpad, ca. 750 B.C. (Appendix I, Chart 4) and the vassal treaties of Esarhaddon, 672 B.C. (Appendix I, Chart 9).

40. See A. Goetze, *Die Annalen des Muršiliš* (MVAG 38/6; Leipzig: J. C. Hinrichs, 1933) 22, 32, 42, 44, 50, etc.

41. The exceptions are the sun-god ($^{d}$UTU) and the goddess Mezulla. The sun-god is not mentioned in the annals, whereas in the treaties he usually holds first place; sometimes, however, the sun-goddess of Arinna holds this position (see G. Kestemont, "Le pantheon des instruments hittites de droit public," *Or* 45 [1976] 156). Although it is generally acknowledged that this goddess and her consort, the storm-god, stood "at the head of the pantheon during the period of the Empire" (*WdM* 173-74), her relation to the sun-god is not clear (O. R. Gurney, *The Hittites* [London: Penguin, 1954] 139). Gurney remarks, "Thus even at the height of the Hittite Empire there was no single unitary hierarchy among the gods" (*Hittite Religion* 6). The second exception is the mention of the goddess Mezulla, daughter of the sun-goddess of Arinna. Although never named in the treaty god-lists, she is regularly named in Mursilis' annals, being his personal deity (see E. Laroche, "Recherches sur les noms des dieux hittites," *RHA* 7 [1946-47] 30).

42. See Chapter IV n. 11.

## Chapter II

1. The text given here is basically that found in *StVA* 3 §528, except that the readings of the Vaticanus manuscript have been followed where

Schmitt reads an emendation. For the text in structured format with emendations and a translation, see Appendix I, Charts 15 and 16.

2. Reiske, *Animadversionum ad Graecos auctores volumen quartum quo Polybii reliquiae pertractantur* (Leipzig: 1763).

3. Ibid. 464-65 (cited from L. F. Benedetto, "Le divinità del giuramento annibalico," *Rivista indo-greco-italica di filologia-lingua-antichità* 3 [1920] 101-2).

4. This imbalance was noted by E. Ziebarth, *De iureiurando in iure Graeco quaestiones* (Göttingen, 1892) 23-24.

5. Schweighäuser, *Polybii Megalopolitani historiarum quidquid superest* (9 vols.; Leipzig: Weidmann, 1792) 6.411.

6. Movers, *Die Phönizier* (3 vols.; Bonn: Weber, 1841) 1.536.

7. Münter, *Religion der Karthager* (Copenhagen: Schubothe, 1821) 105.

8. Pietschmann, *Geschichte der Phonizier* (Berlin: Grote, 1889) 181-82.

9. Meltzer, *Geschichte der Karthager* (3 vols.; Berlin: Weidmann, 1879-1913) 2.145.

10. See, for example, W. W. Baudissin, *Adonis und Esmun: Eine Untersuchung zur Geschichte des Glaubens an Auferstehungsgötter und an Heilgötter* (Leipzig: J. C. Hinrichs, 1911) 283-84.

11. "Baal," in *Ausführliches Lexikon der griechischen und römischen Mythologie* (ed. W. H. Roscher; Leipzig: Teubner, 1884-1937) 1/2.col. 2872.

12. *Encyclopedia Biblica: A Critical Dictionary of the Literary, Political, and Religious History, Archaeology, Geography, and Natural History of the Bible* (ed. T. K. Cheyne and J. S. Black; 4 vols.; New York: Macmillan, 1902) S.V. "Phoenicia."

13. Winckler, *Altorientalische Forschungen*, 1. Reihe (Leipzig: Pfeiffer, 1893-1897) 442-43.

14. "Le panthéon d'Hannibal," *Revue tunisienne* 19 (1912) 329-45; 20 (1913) 29-45, 212-19, 307-14, 447-63, 576-79, 654-57; 21 (1914) 48-55, 164-84. See esp. the first installment.

15. *Revue tunisienne* 19 (1912) 337.

16. Ibid. 343.

17. See n. 3 above.

18. Picard, *Les religions de l'Afrique antique* (Paris: 1954); "Carthage au temps d'Hannibal," *Annuario dell'accademia etrusca di Cortona* 12 (1961-64) 9-36; *Hannibal* (Paris: 1967) 26-35.

19. "Carthage au temps d'Hannibal" 33.

20. Ibid. 34.

21. *Hannibal* 28; "Carthage au temps d'Hannibal" 34.

22. *Hannibal* 29.

23. A. Bounni, "Nabû palmyrénien," *Or* 45 (1976) 48.

24. See J. A. Fitzmyer, "The Phoenician Inscription from Pyrgi," *JAOS* 86 (1966) 288.

25. The concept of a special group of familial gods venerated by the Barcides must be regarded as a creation on the part of Picard. He specifies two deities as members of this group: Melqart, whom he claims the Barcides venerated as their dynastic deity in the Greco-Macedonian fashion ("Carthage au temps d'Hannibal" 32), and Ba$^{c}$al-Šamem, by whom he believes Hannibal swore his oath of eternal enmity toward Rome in 237 B.C. (*Hannibal* 27). But these were major deities at Carthage, clearly high gods of the state religion as well as popular deities. Even if one grants for the sake of argument that these gods were particularly important to the Barcides, Picard fails to demonstrate how their importance to this family was different from their importance to Carthaginians in general.

26. Polybius, *Histories* 3.25.6.

27. "Carthage au temps d'Hannibal" 32.

28. See E. Vassel, "Le panthéon d'Hannibal," *Revue tunisienne* 19 (1912) 336, and P. Xella, "A proposito del giuramento annibalico," *OrAnt* 10 (1971) 190.

29. Bickerman, "Hannibal's Covenant" 19 (cf. Polybius, *Histories* 3.21).

30. Polybius, *Histories* 7.9.1.

31. "International Treaties in Antiquity: The Diplomatic Negotiations Between Hannibal and Philip V of Macedonia," *Classica et Mediaevalia* 15 (1954) 76-77.

32. "Hannibal's Covenant" 7.

33. Walbank, *Historical Commentary* 45.

## Chapter III

1. See Chapter I.

2. Observations on Greek treaties in this chapter are based on the texts in *StVA* 2 and 3.

3. *StVA* 2 §§263, 289; 3 §§481, 492.

4. See Chapter I. The closest parallel from the Greek treaties is *StVA* 3 §481, which is not a treaty but an agreement. The oath of one of

the parties is introduced by *horkos hon ōmosen* followed by the names of the members of that party (cf. *horkos hon etheto Annibas* . . .). But this line comes after nineteen lines of text, in which some of the elements of the agreement are specified, whereas in H/P the introductory section naming the parties comes at the very beginning of the document.

5. J. A. Fitzmyer, *The Aramaic Inscriptions of Sefîre* (BibOr 19; Rome: Pontifical Biblical Institute, 1967); the god-list and a translation are found on pp. 12-13. See also Appendix I, Chart 5.

6. In Sf1, however, a line intervenes between the naming of the contracting parties and the god-list: "And the st[ele with t]his [inscription] he has set up, as well as this treaty. Now (it is) this treaty which Bi-Ga'[yah] has concluded [in the presence of. . .]" (Fitzmyer, ibid.).

7. The text is found in E. Ebeling, "Mittelassyrische Rezepte zur Herstellung von wohlriechenden Salben," *Or* 17 (1948) pl. 31; Ebeling provides a transliteration in *Stiftungen und Vorschriften für assyrische Tempel* (Berlin: Akademie Verlag, 1954) 9. See also Appendix I, Chart 7.

8. Ebeling (*Stiftungen*, ibid.) believes the fragment comes from the middle of the tablet.

9. Wiseman, *Vassal-Treaties*. See also Appendix I, Chart 9.

10. See Chapter I.

11. The text may be found in R. F. Harper, *Assyrian and Babylonian Letters Belonging to the K. Collection of the British Museum* (14 vols.; Chicago: University of Chicago Press, 1892-1914) 11. 1214-17. For a transliteration and translation, see L. Waterman, *Royal Correspondence of the Assyrian Empire: Translated into English, with a Transliteration of the Text and a Commentary* (4 vols.; University of Michigan Humanistic Series 17-20; Ann Arbor: University of Michigan Press, 1930) 2. 266-69. For the god-list itself, see Appendix I, Chart 10.

12. For a transliteration and translation, see K. Deller and S. Parpola, "Ein Vertrag Assurbanipals mit dem arabischen Stamm Qedar," *Or* 37 (1968) 464-66. See also Appendix I, Chart 11.

13. Deller and Parpola (ibid.) estimate about seven lines of text before the beginning of the god-list.

14. For the text and transliteration, see F. V. Scheil, "Sin-šar-iškun fils d'Aššurbanipal," *ZA* 11 (1896) 47-49; a better copy of the text may be found in A. T. Clay, *Babylonian Records in the Library of J. Pierpont Morgan* (4 vols.; New Haven, Conn.: Yale University Press, 1923) Vol. 4, pl. 6, no. 50. R. Borger ("Der Aufstieg des neubabylonischen Reiches," *JCS* 19 [1965] 76) reads the name as dXXX.MU.S[I.SÁ], Sin-šum-lišir. We accept Borger's reading of the name. See also Appendix I, Chart 12.

15. Korošec, *Hethitische Staatsverträge: Ein Beitrag zu ihrer juristischen Wertung* (Leipzig: Weicher, 1931) 14.

16. F6.

17. See A. Kempinski and S. Košak, "Der Išmeriga-Vertrag," *WO* 5 (1970) 191-217.

18. E. von Schuler, *Die Kaškäer* (Berlin: de Gruyter, 1965) 109-27.

19. *StVA* 2 §263; 3 §545.

20. *StVA* 2 §309; 3 §§429, 445, 446, 463, 468, 481, 492, 495, 499, 553.

21. *StVA* 2 §§134, 145, 260, 263, 289, 290, 308; 3 §§429, 446, 463, 476, 481, 492, 495, 499, 545.

22. *StVA* 3 §§495, 545.

23. *StVA* 2 §260; 3 §§429, 446, 463, 476, 481, 492.

24. *StVA* 3 §§429, 463, 468, 481, 492, 499, 553.

25. McCarthy, *Treaty and Covenant* 69 (rev. ed. 109). See also R. Borger, "Marduk-zākir-šumi und der Kodex Hammurapi," *Or* 34 (1965) 168-69.

26. For text, transliteration, and translation, see E. F. Weidner, "Der Staatsvertrag Aššurnirâris VI. [*sic*] von Assyrien mit Mati'ilu von Bît-Agusi," *AfO* 8 (1932-1933) 27-29.

27. The dates of Hammurapi's reign according to the "middle" chronology are 1792-1750; according to the "low" chronology, 1728-1686 (see W. F. Albright, *From the Stone Age to Christianity: Monotheism and the Historical Process* [Garden City, N.Y.: Doubleday, 1957] 10).

28. W. G. Lambert, "The Reign of Nebuchadnezzar I: A Turning Point in the History of Ancient Mesopotamian Religion," *The Seed of Wisdom: Essays in Honour of T. J. Meek* (ed. W. S. McCullough; Toronto: University of Toronto Press, 1964) 6.

29. Ibid.

30. For text, transliteration, and translation, see Weidner, "Staatsvertrag" 17-27; for a recent critical translation, see E. Reiner in *ANET*³ 532-33.

31. The form is the D stative 2nd m. pl. of *tamû*, "to swear" (*AHW* 1318a). McCarthy's translation of this passage, "You call . . . to witness," is misleading (*Treaty and Covenant* 71; rev. ed. 111).

32. For evidence that these are foreign gods, see Appendix II.

33. In the Assyrian treaty the name is written ᵐ*Ma-ti-*ʾ-DINGIR, which is most likely to be read ᵐ*Ma-ti-*ʾ*-il(u)*; in Sf1 the name is written *mtᶜʾl*.

34. E. Dhorme and R. Dussaud, *Les religions de Babylonie et d'Assyrie* [Dhorme]: *Les religions des hittites et des hourrites, des phéniciens et des syriens* [Dussaud] (Les anciennes religions orientales 2; Paris: Presses universitaries de France, 1949) 79.

35. Why are the Sebetti consistently named last among the Mesopotamian oath-gods--so consistently that they constitute the principal caesura in the god-list? First of all, the Sebetti were war-deities, as is clear from their standard epithets, "the warriors without peer" (*qarrād lā šanān*; cf. *Erra* I.8, 18, 23 and *passim*) and "the divine warriors" (*ilāni qardūti*; cf. E/B IV.5; VTE 464). Almost all lists of Assyrian gods in royal inscriptions end with a number of war-gods such as Nergal, Ninurta, Ištar of Arbela, etc. (cf. A/M Rev IV.15-16, 19-20; *ABL* 1105 Rev 17-25; VTE 453-58, 459-60, 464-65). Hence it is understandable that the Sebetti would be included in this latter section as oath-gods. But why at the very end? In official Assyrian god-lists Assyrian deities are always named either individually or in pairs (as in A/M)--never in larger groupings; after the listing of these deities there usually follows a summary phrase invoking the gods in general. Since the Sebetti are by definition "the Seven," it would be logical to mention them after the gods named individually or in pairs and just before the summary phrase. They thus form a kind of bridge between singular (or dual) and plural in the god-list. Then too the Sebetti may be listed in final position because they were actually pictured as marching in last place in Assur's vanguard. One of Sennacherib's inscriptions mentions ten deities who march in front of Assur as he goes into battle. The last-mentioned of these are the Sebetti (see D. D. Luckenbill, *The Annals of Sennacherib* [OIP 2; Chicago: University of Chicago Press, 1924] 141-42). Finally, the fact that the Sebetti consistently hold last place undoubtedly reflects the fact that they were considered "deities of second rank" (J. Bottéro, "Les divinitiés sémitiques anciennes en Mésopotamie," *Le Antiche Divinità Semitiche* [ed. S. Moscati; StSem 1; Rome: Istituto di studi del Vicino Oriente, 1958] 48). In fact they were elevated to the status of "great gods" only in Assyria (not in Babylon), especially in official texts (see L. Cagni, *The Poem of Erra* [SANE 1/3; Malibu, Calif.: Undena, 1977] 19 n. 64). This in turn may reflect a foreign origin. It is possible that the Sebetti were native to Elam (see R. Frankena, *Tākultu: De sakrale maaltijd in het assyrische ritueel: Met een overzicht over de in Assur vereerde goden* [Leiden: Trio, 1953] 111; *WdM* 125). If this is correct, it would further explain why the Sebetti are always mentioned just before the foreign gods in treaties (see Appendix II), forming a transition between Assyrian and foreign deities. On the Sebetti, see S. Graziani, "Note sui Sibitti," *AION* 39 [n.s. 29] (1979) 673-90.

36. Ebeling (*Stiftungen* 9) thinks that *aš-šur* in line 4 of the obverse refers to the god Assur. Not only would it be unique for the supreme god to be listed in this position, but the word is not preceded by the divine determinative. Undoubtedly one must read here $^{d}$XV *aš-šur*[$^{ki?}$-*i-tu*], "the Assyr[ian] Ištar," who appears in a number of official Assyrian god-lists (see, for example, J. N. Postgate, *Neo-Assyrian Royal Grants and Decrees* [Studia Pohl, series major 1; Rome: Pontifical Biblical Institute, 1969] 19, 25, 26, 32, etc.).

37. Assur and Ninlil appear together in A/Q and VTE 414 and 417.

38. The most recent critical edition of this text, with transliteration and translation, is that of R. Borger, *Die Inschriften Asarhaddons Königs von Assyrien* (AfO Beiheft 9; Osnabrück: Biblio-Verlag, 1967 [reprint of the 1956 edition]) §69 (pp. 107-9) and pls. III and IV; for a recent translation, see Reiner, *ANET*$^{3}$ 533-34.

39. R. Frankena, "The Vassal-Treaties of Esarhaddon and the Dating of Deuteronomy," *OTS* 14 (1965) 130.

40. Borger, "Zu den Asarhaddon-Verträgen aus Nimrud," *ZA* 54 [N.F. 20] (1961) 187-96.

41. On the heavenly bodies as witnesses to treaties and the practice of concluding treaties at night, cf. *ABL* 386.16-19:

> UD.15.KÁM *ina kal[-la-ma-ri] ina* ŠÀ *a-de-e*
> *le-e[-ru-bu]* GI$_6$ *šá* UD.15.[KÁM] *ina* IGI
> MUL.MEŠ *liš-ku[-nu]*
>
> On the 15th, at d[awn,] let them en[ter]
> into the treaty; but let them ta[ke] (the
> oath only) on the night of the 15th in
> the presence of the stars.

42. Frankena, "Vassal-Treaties" 127; Weinfeld, "*hšb$^c$t hws'lym šl 'srhdwn mlk 'šwr*" ("The Vassal-Treaties of Esarhaddon--an Annotated Translation"), *Shnaton* 1 (1975) 91 n. 11; Reiner, *ANET*$^3$ 534 n. 4. Note the similarity between *ta[mmu]*, "(DN) is invoked on oath," in VTE and *tummâtunu*, "you are adjured (by)," in M/Š. These are both D statives, the former being the 2nd m. pl. Babylonian form and the latter the 3rd sing. (*tammu*) or pl. (*tammû*) Assyrian form; see Frankena, "Vassal-Treaties" 127.

43. On the identification of the entries here as foreign gods, see Appendix II.

44. "Vassal-Treaties" 132-33. Despite Frankena's claim to the contrary, VTE(D) has no connection with VTE(C), as may be seen from the following considerations. (1) VTE(C) contains the Sebetti (464), who, as we have seen, conclude the listing of Assyrian gods. The list is further concluded in its entirety by the summary phrase and accompanying curses, which, like all other entries in (C), is marked off by ruled lines. Frankena is wrong in thinking that 472-93 is a separate section by itself; it is the conclusion to (C), since this kind of summary phrase both in treaties and other types of official documents serves to conclude a list of deities. (2) After (C) comes an entirely different section (494-512), containing "a solemn statement of the vassals, . . . reaffirm[ing] their obedience to the stipulations of the treaty" (ibid. 132). (3) The fact that Assur is named again (518) is a clear indication of the beginning of a new god-list. (4) There is no example in Neo-Assyrian royal documents of a list "interrupted" (as Frankena claims here) by a long digression and then "resumed."

45. Ibid. 133.

46. E.g., Šamaš in 545 and 649.

47. See Chapter I.

48. See Appendix II.

49. See Appendix I, n. 11.

50. R. Borger, in a review of *CAD* "Z" (*BO* 22 [1965] 167), mentions a communication from R. Frankena: "Wiseman, *Treaties*, 1. 476f, is a quotation from the Codex Hammurapi (Rev XXVII 34-40). . . ." In S/N too the mention of Sin after a general curse of "the great gods" may be due to

patterning on the order of curses in CH or a later series of curses based on the latter. In S/N the following curse is associated with the great gods: "Above among the l[iving may they uproot th]em, below in the netherworld may they cause [their] g[hosts] to thirst for [water]." In CH this same curse (Rev XXVII 34-40) is followed immediately by the name of Sin and curses associated with him.

51. The text of H/P does not have *enantion* before *kai Hēliou kai Selēnēs kai Gēs*. But see Chapter IV.

52. Bickerman, "Oath of Hannibal" 92.

53. *ABL* 202 Rev 5-7. The translation is from A. L. Oppenheim, *Letters From Mesopotamia* (Chicago/London: University of Chicago Press, 1967) 155.

54. Bickerman ("Oath of Hannibal" 92) notes that *enantion* was used "in a forensic sense" in Greek.

55. For the reading of IGI as *pān* before the names of witnesses in legal documents, see J. N. Postgate, *Fifty Neo-Assyrian Legal Documents* (Warminster: Aris and Phillips, 1976) 9. For the reading of this sign as *maḫar* in the same context, see R. Borger, *Babylonisch-Assyrische Zeichenliste* (AOAT 33; Neukirchen-Vluyn: Neukirchener Verlag, 1978) 172.

56. Fitzmyer, *Sefîre* 33.

57. Ibid.

58. No consensus has been reached as to the location of the land of KTK (see Fitzmyer, ibid. 127-35); but Fitzmyer (ibid. 2) speaks of Bir-Ga'yah, king of KTK, as "a powerful Mesopotamian overlord." If this is correct, there would be no problem in supposing that KTK had at some earlier period adopted the Babylonian pantheon as its own, yet saving the highest position in the pantheon for its native supreme deities; Assyria had done precisely the same thing, elevating Assur (the city-god of Assur) to the supreme position in the pantheon. On the identification of KTK, see most recently N. Na'aman, "Looking for KTK," *MIO* 9 (1977-78) 220-39.

59. Fitzmyer, *Sefîre* 36.

60. *KAI* 2.245-46.

61. Gibson, *Textbook of Syrian Semitic Inscriptions: Volume II: Aramaic Inscriptions, Including Inscriptions of the Dialect of Zenjirli* (Oxford: Clarendon Press, 1975) 36.

62. Fitzmyer, *Sefîre* 36.

63. Jastrow, *Die Religion Babyloniens und Assyriens* (2 vols.; Giessen: Töpelmann, 1905) 1.185.

64. But in Sf1 the phrase "all the gods of RḤBH and 'DM" does not terminate the list, as we shall show below. Yet in some way this expression does appear to mark a "seam" in the god-list (see n. 67 below). It is interesting to note that in the "pantheon list(s)" from Ugarit (see Chapter IV n. 92) *ddmš* comes just before the summary phrase *pḫr 'ilm*, "the assembly

of the gods" (see J. Nougayrol in *Ugaritica V: Nouveaux Textes accadiens, hourrites et ugaritiques des archives et bibliothèques privées d'Ugarit: Commentaires des textes historiques (Première partie)* [ed. J. Nougayrol et al.; MRS 16; Paris: Imprimerie nationale, 1968] 45; E. Laroche, ibid. 521); like Nin-karra(k), the Hurrian *ddmš* was a goddess of healing (ibid. 57).

65. Fitzmyer, *Sefîre* 33; Gibson, *Textbook* 37.

66. The summary phrase in E/B IV.8-9 *seems* to divide the god-list into Assyrian and non-Assyrian deities; but see Appendix II.

67. The line "all the gods of RHBH and ʼDM" probably contains two geographical names (Fitzmyer, *Sefîre* 36); we would take these to be sub-areas of KTK. The phrase itself quite possibly marks a *minor* division, i.e., in the first part of the god-list, dividing the specifically Babylonian from the Assyrian components of the list:

> (1) The entries from Marduk through Nikkar constitute the Babylonian portion of the list. Note first that Marduk, the chief god of the Babylonian pantheon, is named in first place with his consort, followed by his son Nabu with his consort. By contrast Assur, the chief god of Assyria, is not named at all. Secondly, Nin-karra(k) is the form of the healing goddess' name most commonly found in Babylonia; in Assyria the name Gula is more common (cf. A/M Rev IV.11).
> (2) The entries Hadad of Aleppo and the Sebetti are most likely from Assyrian treaty tradition. Hadad of Aleppo was invoked as an oath-god in northern Syria, Hatti, Assyria, and (probably) Ugarit (see n. 69). All of these lands are neighboring to Aleppo, whereas Babylonia is rather distant. This god is not found in official Babylonian god-lists but is named in A/M Rev IV.18. Secondly, the Sebetti are found in official god-lists from Assyria but not Babylonia (see n. 35).

68. The restoration [*wqdm hdd ḥ*]*lb* is supported by all commentators on this treaty: see A. Dupont-Sommer and J. Starcky, *Les inscriptions araméens de Sfiré (Stèles I et II)* (Paris: Imprimerie nationale, 1958) 34; Fitzmyer, *Sefîre* 36; *KAI* 2.246; Gibson, *Textbook* 29.

69. At first sight the inclusion of Hadad of Aleppo in a list of Mesopotamian gods seems strange (cf. A/M Rev IV.18). But the significance of this god extended far beyond that of a local Syrian deity. In addition to appearing as an oath-god in A/M, he was venerated by Šalmaneser III, who offered a sacrifice to him before the battle of Qarqar (see H. Klengel, "Der Wettergott von Ḫalab," *JCS* 19 [1965] 92). He is also listed in the "Address-book of the Gods" (Frankena, *Tākultu* 124.116), indicating that his cult in Assyria continued at least till the time of Sennacherib (ibid. 122). Hadad of Aleppo appears as an oath-god in treaties from Alalakh (*AT* 2.77; 3.46; 126.27; in these texts $^{d}$IM = Hadad of Aleppo [see Klengel, "Wettergott" 89]). He was an important deity in Hatti as well and was commonly invoked as an oath-god in Hittite treaties (ibid. 91). In the Egyptian version of the treaty between Ramses II and Hattusilis III he is invoked as a divine witness under the name *św*t*ḫ ḫrp*, "Seth [i.e., the storm-god] of Aleppo" (ibid. 88). At Ugarit sacrifices were offered to *b*$^{c}$*l ḫlb* (609.2.4),

equivalent to *t̲t̲*[*b*.]*ḫlbǵ* (*CTA* 166.10), "Tešub [the Hurrian storm-god] of Aleppo." Interestingly he is named in one god-list from Ugarit which is clearly an official list and perhaps the prototype of a treaty god-list (see *Ugaritica V* 321 and Chapter IV n. 92). Therefore Hadad of Aleppo must not be considered an intrusive "Canaanite" entry into what is otherwise a list of Mesopotamian oath-gods. He clearly belonged to the treaty traditions of a wide area of the ancient Near East from the second to the middle of the first millennium B.C.

70. Fitzmyer, *Sefîre* 37.

71. The healing-goddess *nkr* (Nikkar) appears in A/M under her Assyrian name, Gula.

72. *Pace* Fitzmyer, ibid. 33; Gibson, *Textbook* 36. If this were such a list we would expect the chief deity of Arpad to be named first. But in the lines listing the foreign gods (i.e., the gods of Mati'ilu and hence the gods of Arpad) in A/M (see Appendix II) the first-named deity is apparently Dagan followed by a city-name. Nothing resembling 'El and/or ᶜElyon is found there.

73. "Thus *'l wᶜlyn* seems to mark the transition in the list from the cultically important deities to those deities who, apart from their role as witnesses, seldom occur outside cosmogonies" (J.J.M. Roberts, "The Davidic Origin of the Zion Tradition," *JBL* 92 [1973] 332 n. 16).

74. Fitzmyer, ibid. 37. See also F. M. Cross, *Canaanite Myth and Hebrew Epic: Essays in the History and Religion of Israel* (Cambridge: Harvard University Press, 1973) 51, who concedes, "It is even possible to interpret the pair as a double name of a single god. . . ."

75. According to K. Tallqvist, *Akkadische Götterepitheta* (StOrFen 7; Leipzig: Harrassowitz, 1938), the epithet *nūru* ( = *nr*) is frequently applied to Šamaš (133-34, 456); but he gives no indication that Aya ever bore this title. Note too that at Ugarit *špš*, the Sun-goddess, bore the same epithet (with the feminine ending) in the stock phrase *nrt 'ilm špš* (*CTA* 6[62].1.13, etc.), "Šapaš, the luminary of the gods."

76. See most recently C. E. L'Heureux, *Rank Among the Canaanite Gods El, Baᶜal, and the Repha'im* (HSM 21; Missoula, Mont.: Scholars Press, 1979) 46.

77. See also M. C. Astour, "Two Ugaritic Serpent Charms," *JNES* 27 (1968) 28-29.

78. Fitzmyer, *Sefîre* 37; Dupont-Sommer and Starcky, *Sfîré* 34; Gibson, *Textbook* 37.

79. This connection was, to my knowledge, first noted by D. R. Hillers ("Hear, O Heaven, and Give Ear, O Earth" [Schaff Lectures, No. 3 (unpublished paper)] 21).

80. See p. 90 and nn. 443-44 below.

81. See Cross, *Canaanite Myth* 51. Note the phrase "'El, creator of the earth," which appears in a Hittite version of a West Semitic myth

(*elkunirša* < *ʾēl qōnê ʾarṣ*; see H. Otten, "Ein kanaanäischer Mythus aus Boğazköy," *MIO* 1 [1953] 125-50; H. G. Güterbock, "Hittite Mythology," *Mythologies of the Ancient World* [ed. S. N. Kramer; Garden City, N.Y.: Doubleday, 1961] 155); note also *ʾl qn ʾrṣ* in Canaanite texts (*KAI* 26 A III.18; 129.1) and *ʾl qn ʾr*$^{c}$ in Palmyrene texts (see R. Dussaud, *La pénétration des arabes en Syrie avant l'Islam* [Paris: Geuthner, 1955] 169 and n. 2).

82. See, for example, Frankena, *Tākultu* 7.VI.4 ([heavens and] earth), VIII.3 (mountains, [rivers]).

83. F. M. Cross, "The 'Olden Gods' in Ancient Near Eastern Creation Myths," *Magnalia Dei, the Mighty Acts of God: Essays on the Bible and Archaeology in Memory of G. Ernest Wright* (ed. F. M. Cross, W. E. Lemke, and P. D. Miller, Jr.; Garden City, N.Y.: Doubleday, 1976) 332.

84. See O. R. Gurney, "The Hittite Prayers of Mursili II," *AAA* 27 (1940) 81; E. O. Forrer, "Eine Geschichte des Götterkönigtumus aus dem Hatti-Reiche," *Mélanges Franz Cumont* (AIPHOS 4; Brussels: Sécretariat de l'Institut, 1936) 698.

85. E. Laroche, "Les dénominations des dieux 'antiques' dans les textes hittites," *Anatolian Studies Presented to Hans Gustav Güterbock on the Occasion of his 65th Birthday* (ed. K. Bittel et al.; Istanbul: Nederlands historisch-archaeologisch Instituut in het Nabije Oosten, 1974) 175-85.

86. See D. O. Edzard, *WdM* 42. This equivalence has been established by H. Otten, "Eine Beschwörung der Unterirdischen aus Boğazköy," *ZA* 54 [N.F. 20] (1961) 114-57, esp. 115, 157.

87. See Kestemont, "Pantheon" 153.

88. Cross, however, seems to understand these natural phenomena as part of the old(en) gods; see "Olden Gods" 334.

89. However, we shall retain the traditional designation "divinized" with respect to these natural phenomena, since they do form part of a list of divine witnesses.

90. After W3 Rev IV.43; F4 IV.23; KUB XXIII 77a Obv 8.

91. KBo VIII 35 Obv II.11-12; see E. von Schuler, *Kaškäer* 117. For an English translation, see p. 35 below.

92. See Edzard, *WdM* 60. It should be noted, however, that Mesopotamian religion did not have *a* creator god par excellence corresponding precisely to ʾEl.

93. A. Goetze, *ANET*$^{3}$ 125a: "Let them [i.e., the olden gods] bring [for]th the olden copper *knife* with which they severed heaven and earth." Does "they" here refer to the olden gods, or is it to be understood in a general sense (like French *on*, German *man*)? E. von Schuler (*WdM* 205) seems to understand the passage in the latter sense.

94. Von Schuler, *WdM* 185.

95. See *Ugaritica V* 246.35´´:

[$^{d}$*En-líl*        *K*]*u-*[*m*]*ur-wi*        *Ilum*(DINGIR)$^{lum}$

The restoration of $^{d}$En-líl in the first column is beyond question, since the list follows the order of the list in *Ugaritica V* 212. Thus Sumero-Akkadian An, Antum, Enlil, Ninlil = Hurrian Ani, Ašte-Ani-wi, Kumurwi [ = Kumarbi], Ašte-Kumurwi-ni-wi = Ugaritic Šamūma, Ta(h)amatum, Ilum [ = 'El], Aširtum [ = 'Ašerah]. However, one must remember that such "identifications" were never absolute nor exhaustive. Kumarbi seems to be equated also with Dagan at Ugarit (ibid. 524).

96. F4 B Rev IV.25; F5 IV.23; KUB XL 42 Rev 7; XL 43 Rev 4.

97. McCarthy, *Treaty and Covenant* 71, claimed that "unlike the Hittite[s] they [i.e., the Assyrians] often ignored foreign gods in their treaties." This would imply that the inclusion of foreign gods in Hittite treaties would be the rule and not the exception, which is not true. But in the revised edition of his book (111) McCarthy has deleted this sentence.

98. W1 Rev 54-58; W2 Rev 40-43; KUB XXIII 77a Obv 12-18.

99. Fitzmyer, *Sefîre* 13; Dupont-Sommer and Starcky, *Sfiré* 17; *KAI* 2.239; Gibson, *Textbook* 28.

100. The phrase is interpreted as declarative by Fitzmyer, *Sefîre* 13; Dupont-Sommer and Starcky, *Sfiré* 17; H. Donner and W. Röllig, *KAI* 2.239; as imperative, however, by Gibson, *Textbook* 29.

101. According to Wiseman's interpretation, VTE 23b-24 is to be translated: "[the gods] have affirmed, have laid on, (and) made (this treaty)"; he reads the text here *ú-dan-nin-*[*ú*] *is-ba-tú iš-ku-nu-n*[*i*] (*Vassal-Treaties* 32). From this one would be led to think that VTE contains a statement of divine ratification or attestation similar to that at the end of Sf1. But prescinding from the incorrect reading *is-ba-tú* (see Borger, "Asarhaddon-Verträgen" 175; Frankena, "Vassal-Treaties" 126 n. 2), the verbs are subjunctive, so that Wiseman's translation cannot be right. The subject here is Esarhaddon, not the gods (see Frankena, "Vassal-Treaties" 125-26; Weinfeld, "*hšb$^{c}$t*" 89-90 n. 5; Reiner, *ANET*$^{3}$ 534).

102. See, for example, KBo IV 10 Rev 50b-51:

50b *nu ka-a-ša a-pé-e-da-ni me-mi-ni LI-IM*
DINGIR.MEŠ *tu-li-ya ḫal-zi-ya-an-te-eš*
51 *nu uš-kán-du iš-ta-ma-aš-kán-du-ya*
*na-at ku-ut-ru-e-eš a-ša-an-du*

Now behold, the thousand gods are called
to assembly at (the conclusion of) this *treaty*
[or perhaps "on this occasion"];
and let them *look upon* and listen to
(it), and let them be *witnesses* (to it).

This passage is found just before the list of gods proper. Compare, however, ABoT 56 II.15´-16´, where a passage almost identical to line 51 above occurs at the *end* of the list:

15´ *na-at ke-e-d[a-ni li-in-ki-ya ku-ut-ru-eš a-ša-an-du]*
16´ *nu uš-kán-du iš-da[m-ma-aš-kán-du-ya]*

And may they [be *witnesses*] at (the conclusion of) th[is *treaty*],
And may they *look upon* and lis[ten to (it)].

As in Sf1 the gods are specifically called to be *witnesses* to (*kutrueš ašandu*) and to *look upon* (*uškandu*) the *treaty* (*linkiya*); the invocations in Sf1 and ABoT 56 II follow the same order, and both occur at the end of the list of divine witnesses (for the restoration of ABoT 56 II.15´, see n. 160 below). Cf. also VTE 494: DINGIR.MEŠ *an-nu-te lid-gu-lu* . . . , "May these gods *see/take note*. . . ."

103. Parallels between Sf1 and H/P have been noted by Dupont-Sommer and Starcky, *Sfiré* (passim).

104. See n. 51 above.

105. "Panthéon." See Appendix I, Chart 14.

106. Ibid. 147.

107. Ibid. 150.

108. Ibid. 151.

109. The deities in question are dU NIR.GÁL, dU Piḫaššašši, dU ŠÁR. ŠÁR, and Lulutašši.

110. Kestemont, "Panthéon" 157. On the relationship of Šēri and Ḫurri to the Storm-god, see *WdM* 195.

111. Some scholars hold that these are all manifestations of the same god, others that they have become distinct deities. The former opinion is supported by Güterbock, "Hittite Religion," *Ancient Religions* (originally published as *Forgotten Religions*) (ed. V. Ferm; New York: Philosophical Library, 1950) 88; the latter by von Schuler, *WdM* 208.

112. It is true that the "canonical" Syro-Anatolian list of deities has a section of high gods followed by a section naming the corresponding consort goddess (E. Laroche, "Panthéon national et panthéons locaux chez les Hourrites," *Or* 45 [1976] 97-98); a similar arrangement appears in VTE(A). But this schema does not correspond to the structure of the god-list in the Hittite treaties.

113. See n. 127 below.

114. Even Ištar here may be conceived as masculine, in her role as warrior; at least this is the opinion of Laroche ("Panthéon national" 97).

115. "Panthéon" 153.

116. Ibid.

117. Akkadian *gabbašunu/gabbišunu* (W1 Rev 15; W3 Rev IV.38-40; W4 Rev 5-6; RS 17.349B Rev 11´); Hittite *ḫumanteš/ḫumanduš* (F4 B Rev IV.20 [see n. 118]; F6 I.57; KBo XII 31 Rev IV.8´-9´; KUB XXVI 39 Rev IV.23-24; XL 42 Rev 10; AboT 56 II.26.

118. W3 Rev IV.41; F1 IV.15; F4 B Rev IV.22. It is found once (F4 B III.19-20) with "the gods (of the?) Lulaḫḫi, the gods (of the?) Ḫapiri." Here, however there has been a displacement in the text. *Ḫu-u-ma-an-te-eš ŠA* KUR $^{\mathrm{uru}}$*Ḫat-ti* is placed after the Lulaḫḫi and Ḫapiri only in this text, whereas the phrase regularly goes with the summary expression (see the passages listed in the previous note.

119. F6 I.52-53.

120. After F4 B III.20; KUB XXVI 39 Rev IV.22; XL 42 Rev 9.

121. On this equivalence see Laroche, "Dieux antiques" 176; H. M. Kümmel, "Die Religion der Hethiter," *Theologie und Wissenschaft* (ed. U. Mann; Darmstadt: Wissenschaftliche Buchgesellschaft, 1973) 72.

122. Before: W3 Rev IV.37; F1 IV.13; KBo IV 10 Rev 3; after: W1 B Rev 28; W2 Rev 22; F4 B Rev IV.21; ABoT 56 II.27.

123. Kestemont observes, "this deity [i.e., the sun-goddess of the netherworld] is here displaced and placed in the middle of group II E" ("Panthéon" 167 n.). See also Gurney, *Hittite Religion* 5 n. 6.

124. Kümmel, "Religion der Hethiter" 72.

125. After W3 IV.18; F4 B Rev IV.6; F6 I.47; RS 17.342 Rev 3´; 17.450 Rev 4´; KUB XXVI 39 Rev IV.10.

126. Note that in Sf1 the supreme gods (DN and *mlš*, Marduk, Nabu, and consorts) are followed by Erra and Nergal (with their consorts), who are also war-gods.

127. In Akkadian the logogram $^{\mathrm{d}}$LAMMA usually stands for *lamassu*, "protective deity," and only occasionally designates a specific god (such as Papsukkal; see *CAD* "L" 64a). In Hittite, however, the logogram not only stands for the common noun *annari-* ( = *lamassu*) but regularly for several deities: Innara, the "stag-god" (see Gurney, *Hittite Religion* 8 n. 6), or N/Lubadig (see H. Otten, "Die Götter Nupatik, Pirinkir, Ḫešue und Ḫatni-Pišaišaphi in den hethitischen Felsreliefs von Yazilikaya," *Anatolia* 4 [1959] 29-30). Thus in formulas like $^{\mathrm{d}}$LAMMA $^{\mathrm{uru}}$*Ḫatti*, should one read "the protective deity of Hatti," "Innara (f.) of Hatti," or perhaps "N/Lubadig (m.) of Hatti"?

128. *WdM* 202.

129. Ibid. 201. But see Gurney, *Hittite Religion* 12.

130. W1 Rev 45-46; W2 Rev 17-18; W3 Rev IV.23-24; F4 B Rev IV.10.

131. W1 Rev 46.

132. "Panthéon" 150.

133. E.g., in F4 B Rev IV.10, 13; F5 IV.12, 14.

134. Gurney, *The Hittites* 135.

135. E.g., F5 IV.9; F6 I.47.

136. Laroche, "Panthéon national" 97.

137. W3 Rev IV.27; F4 B Rev IV.12; F5 IV.13; ABoT 56 II.19.

138. After W3 Rev IV.24; KUB XXVI 39 Rev IV.15; RS 17.342 Rev 7´; 17.351A Rev 6´.

139. After F4 B Rev IV.13.

140. Otten, "Yazilikaya" 33.

141. F6 I.52-53.

142. Kestemont, however, includes these gods in a separate section ("Panthéon" 163).

143. Kestemont's Chart 12 (ibid.) is misleading. I cannot see where he finds Marduk ($^{d}$AMAR.UTU) or "the gods of the army" (DINGIR.MEŠ KARAŠ or *tuzziyaš* DINGIR.MEŠ) in KUB XXVI Rev IV.

144. F6 I.53.

145. F6 I.53.

146. KUB XXVI 39 Rev IV.18.

147. After W3 Rev IV.29; F1 IV.3; F5 IV.16; RS 17.349B Rev.3´.

148. Laroche, "Recherches" 7-139.

149. "Panthéon" 151; Gurney, *Hittite Religion* 5.

150. Kestemont ("Le traité entre Mursil II de Hatti et Niqmepa d'Ugarit," *UF* 6 [1974] 117 n. 215) claims that *lulaḫḫi* and *ḫapiri* are divine names, not gentilics, since the terms are in the plural in Hittite. But actually the number of the noun is not the important factor, since it would be plural whether one takes Lulaḫḫi as a group of gods or an ethnic designation. What is important here would seem to be the case of the proper nouns in Hittite. But the evidence is ambiguous. Forms like *Lu-la-ḫe-e-eš* (F4 B Rev IV.19; KUB XXIII 77a Obv.8) must be nom. or acc. pl. (see J. Friedrich, *Hethitisches Elementarbuch: 1. Teil, Kurzgefasste Grammatik* [Heidelberg: Winter, 1974] §63]) whereas forms such as *Lu-la-ḫi-ia-aš* (KUB XL 42 Rev 3; XL 43 Rev 1) could be nom. or gen. pl. (see Friedrich, ibid. §§62-63). One text (KBo IV 10 Rev 3) places the DINGIR sign before Lulaḫḫi and Ḫapiri, which would support Kestemont's position. But another (W2 Rev.21) reads DINGIR.MEŠ *ša Nu-la-aḫ*[*-ḫi-i*], where the Akkadian particle *ša* shows that the noun must be parsed as genitive. Probably the Hittite scribes themselves were not clear on the matter. On the other hand, it is hard to see how Kestemont can flatly deny that these terms designated ethnic groups. Gurney cites a text which speaks of "Lulahhi men, who have a

dagger in the belt, draw bows and hold arrows" (*Hittite Religion* 29-30). Here the Lulaḫḫi are clearly a warlike group of people, not gods. At least the matter seems to be more complicated than Kestemont supposes. See also M. Weippert, *Die Landnahme der israelitischen Stämme in der neueren wissenschaftlichen Diskussion* (FRLANT 92; Göttingen: Vandenhoeck & Ruprecht, 1967) 73 n. 3; and *CHD* 3/1, 79-80.

151. M. Greenberg, *The Ḫab/piru* (AOS 39; New Haven, Conn.: American Oriental Society, 1955) 77.

152. Ibid.

153. M. Liverani, *Storia di Ugarit nell'età degli archivi politici* (StSem 6; Rome: Instituto di studi del Vicino Oriente; Università di Roma, 1962) 89.

154. Liverani, "La preistoria dell'epitheto 'Yahweh ṣĕbā'ōt'," *AION* 17 (1967) 332.

155. *Hittite Religion* 5.

156. W1 Rev 58-59; W2 Rev 25; W3 Rev IV.45-47; W4 Rev 10-11; F1 IV. 19-20; RS 17.338 Rev 5´; KBo VIII 35 Obv II.13; KUB XXIII 77a Rev 10.

157. W1 Rev 38-39; W2 Rev 10-11; F2 Rev 55-56; RS 17.79 + 374 Rev 54´; KUB XXVI 41 Obv 5; XXIII 77a Obv 2-3.

158. After W4 Rev 11; F1 IV.20; RS 17.338 Rev 5´; KBo VIII 35 Obv II.13; KUB XXIII 77a Rev 10.

159. Only one treaty (W2) has a line immediately before the summons, separating it from the god-list.

160. Akkadian *ana annî riksi u (ana) māmīti lū šībūtu* (W3 Rev IV.45; W4 Rev 10-11; RS 17.338 Rev 5´; F1 IV.19-20); Hittite *nat kēdani linkiya kutru(en)eš ašandu* (KBo IV 10 Rev 51 [omits *kēdani linkiya*]; VIII 35 Obv II.13; XII 31 Rev IV.15´-16´[*lingai* for *linkiya*]; KUB XXIII 77a Obv 10; XXVI Obv I.6; XXVI 41 Obv 4-5; ABoT 56 II.31-32).

161. Attested only in the Hittite treaties in Akkadian: *lizzizzū* (W1 Rev 39,58; W2 Rev 11).

162. Akkadian *liltemû* (W1 Rev 39,59; W2 Rev 11,25); Hittite *ištamaškandu* (KUB XXIII 77a Obv 11; ABoT 56 II.32; in a summons before the god-list, F5 III.83).

163. See n. 102 above.

164. "Dieux antiques" 177.

165. Von Schuler, *Kaškäer* 127.

166. *Hittite Religion* 7 n. 1.

167. Translation in von Schuler, *Kaškäer* 110; Hittite in transliteration and translation in Laroche, "Dieux antiques" 177.

168. It is somewhat difficult to decide whether $^{d}$UTU in this and other short lists is to be read as the sun-*god* (i.e., the sun-god of heaven) or the sun-*goddess* (i.e., the sun-goddess of Arinna). One short list, which is only partially preserved (KUB XXVI 41 = the Išmeriga treaty; see Kempinski and Košak, "Išmeriga" 192), mentions both sun-deities (Obv 9): $^{d}$UTU $^{d}$IŠKUR $^{d}$UTU $^{ur}$[$^{u}$*A-ri-in-na*?]. The "Ritual Before Battle" (KUB XXXI 146, in von Schuler, *Kaškäer* 168 [translation and transliteration]; see A. Goetze, *ANET*$^{3}$ 354), which is not actually a treaty but closely related to that literary genre, clearly lists the feminine deity (2-3): *A-NA* $^{d}$UTU $^{uru}$*A-ri-in-na Ù A-NA* $^{d}$IŠKUR. But in the Kaškean treaties von Schuler (*Kaškäer* 110, 117) consistently reads $^{d}$UTU as "der Sonnengott."

169. $^{d}$LAMMA comes immediately after the supreme gods in the short list KUB XXIII 77a Obv 3 and XXXI 146 Obv I.3.

170. Both Laroche ("Dieux antiques" 177) and von Schuler (*Kaškäer* 115) understand the phrase $^{d}$*Iš-ḫa-ra li-in-ki-aš iš-ḫa-a-aš* to consist of Išḫara plus an epithet, although the latter is masculine ("lord"). One wonders, however, whether in this case *linkiyaš išḫāš* could not refer rather to the moon-god, since this is his usual epithet in this section (W1 Rev 46: $^{d}$XXX EN *ma-mi-ti*; W3 Rev 18: [$^{d}$XXX] EN *ma-mi-ti*; F5 IV.12: $^{d}$XXX! EN *NI-EŠ* DINGIR-*LIM*). Thus although he is usually named first in this section (at least in the later, standard treaties), perhaps one could translate: "Ištar, Išḫara, (and) <the moon-god>, the lord of the oath."

171. The Telepinu cycle (Section III) is preserved in only one of the short lists, KUB XXXI 146 Obv I.4 (also 27); perhaps it never existed in the other (damaged) lists.

172. See H. Otten, "Die Gottheit Lelvani der Boğazköy-Texte," *JCS* 4 (1950) 120. This phrase serves as a final (or penultimate?) statement in KUB XXVI 41 Obv 11 (Išmeriga treaty): *ne-pí-ša-aš ták-na-aš-ša* DINGIR.M[EŠ].

173. Laroche ("Dieux antiques" 178) notes that the position of the phrase here is "abnormal"; the phrase olden gods occupies the place (just before the summary phrase) sometimes occupied by Ereškigal (as, for example, in W3 Rev IV.41 and F1 IV.14-15). In W1 $^{d}$*Er*[*eš-k*]*i-gal* occurs (in the variant Bo 1578) as a variant reading to DINGIR.MEŠ *er-ṣe-ti* (see Weidner, *PDK* 30 n. n).

174. One entry from Section VI is found in the short list KUB XXVI 41 Obv 9, and several from this section appear in the short list KUB XXIII 77a Obv 6-7; the latter text also mentions "the gods (of the?) Lulaḫḫi, the gods (of the?) Ḫapiri" (8), our Section VII B. For Section III, see n. 171 above.

175. Von Schuler, *Kaškäer* 117.

176. This also happens in the case of several of the short god-lists: KUB XXVI 41 Obv 11; XXXI 146 Obv 5-6, 28.

177. See p. 94.

178. See Chapter III and n. 117 above.

179. F6 I.52-53.

180. See Chapter III.

181. In some of the Hittite treaties, $^{d}$LAMMA $^{uru}$*Hatti*, "the protective deity of Hatti," heads this section rather than $^{d}$LAMMA (W1 Rev 43; W2 Rev 15; RS 17.450 Rev 5´); the former would correspond more closely with *daimōn Karchēdoniōn*, "the protective deity of the Carthaginians," which heads the second line in the H/P god-list.

## Chapter IV

1. See Chapter II.

2. *Archaeology and the Religion of Israel* (Baltimore: Johns Hopkins Press, 1968) 74.

3. Caquot, "Problèmes d'histoire religieuse," *La Siria nel Tardo Bronzo* (ed. M. Liverani; Orientis Antiqui Collectio 9; Rome: Centro per le antichità e la storia dell' arte del Vicino Oriente, 1969) 62.

4. "Ba$^{c}$al Samēm and ʾĒl," *CBQ* 39 (1977) 470.

5. Ibid. 471.

6. Ibid.

7. Weidner, "Staatsvertrag" 22.

8. See provisionally G. Pettinato, "The Royal Archives of Tell Mardikh-Ebla," *BA* 39 (1976) 48.

9. See p. 56.

10. See pp. 50-56.

11. Cross (*Canaanite Myth* 10 n. 32) and L'Heureux, (*Rank Among the Canaanite Gods* 68) have argued that the order of gods here does not reflect hierarchy but a chronological series of patron deities--at least the first three entries. Cross is surely correct in his argument that the gods mentioned in *KAI* 24.15-16 (*b*$^{c}$*l.ṣmd.ʾš.lgbr* . . . *b*$^{c}$*l.ḥmn.ʾš.lbmh.wrkbʾl.b*$^{c}$*l.bt*), from the ninth century B.C. are the same as *ḥdd* . . . *ʾl* . . . *rkbʾl.b*$^{c}$*l.bt* in *KAI* 214.2, 11, 18; 215.22 from the eighth century B.C. Since in *KAI* 24 Kilamuwa lists his predecessors (not including his father *ḥyʾ* and his brother *šʾl*) as *gbr* and *bmh*, in that order (2-3), Cross reasons that the sequence "Ba$^{c}$al-ṢMD, who belongs to Gabber . . . (and) Ba$^{c}$al-Ḥamon, who belongs to BMH" is based on chronological considerations. Hence he concludes, "There is no doubt possible concerning El's place at the head of the pantheon" (*Canaanite Myth*, ibid.).

But doubt is indeed possible. Cross and L'Heureux fail to take into account the fact that the Canaanite pantheon was not native to extreme northern Syria. Šamʾal lay in the vicinity of Kizzuwatna, dominated by Hurrian influence in the second millennium. In fact it was from Kizzuwatna that Hattusilis III imported the Hurrian pantheon into the state religion of Hatti in the thirteenth century (see Gurney, *Hittite Religion* 17). This raises the possibility that the sequence of gods in the Šamʾal inscriptions--regardless of their relationship to Gabber and BMH--might reflect the canonical order of deities in the Hurrian pantheon. A comparison of the seven great gods at the head of the "canonical Syro-Anatolian" pantheon (Laroche, "Panthéon national" 96) with the sequence of the deities named in *KAI* 214.2-3,11 makes this absolutely clear:

| | | | | |
|---|---|---|---|---|
| 1. | Tešub (storm-god) | = | 1. | Hadad |
| 2. | Kumarbi | = | 2. | ʾEl |
| 3. | Eya (Ea) | | | |
| 4. | Kušuḫ (moon-god) | = | 3. | Rakiḇ-ʾel (moon-god) |
| 5. | Šimegi (sun-god) | = | 4. | Šamaš |
| 6. | Aštabi | | | |
| 7. | Nubadig | = | 5. | Rešep (*ʾrqršp* in 214.11) |

(On the equation of Kumarbi and ʾEl [at Ugarit], see Chapter III n.95. Note that Cross himself argues for the identification of Rakib-el as the moon-god, Yariḫ [*Canaanite Myth* 10 n.32]. On the identification of Nubadig with Rĕšep [at Ugarit], see my article "dLAMMA and Rešep at Ugarit: The Hittite Connection," *JAOS* 98 [1978] 466-67.)

12. For Aleppo, see Klengel, "Der Wettergott von Ḫalab" 90; for Alalakh, D. J. Wiseman, *The Alalakh Tablets* (Occasional Publications of the British Institute of Archaeology at Ankara 2; London: British Institute of Archaeology at Ankara, 1953) 16.

13. *Canaanite Myth* 32.

14. "ʾēl," *TDOT* 1.253.

15. I fail to understand why Cross insists that ʾElat is the principal wife of ʾEl (Kronos) in Sakkunyaton, unless this is an a priori conclusion. Cross equates Dione in Sakkunyaton (*Pr. Ev.* 1.10.35) with ʾElat (*Canaanite Myth* 28-29 n.90). But besides Dione ʾEl also took Rhea and Astarte to wife (1.10.22-24); in fact Sakkunyaton explicitly calls Astarte *hē megistē* (1.10.31), "the greatest [of the goddesses]," whereas he gives no such epithet to Dione.

16. See pp. 60-61.

17. O. Eissfeldt, "Baʿalšamēm und Jahwe," *ZAW* 57 (1939) 4.

18. Ibid.

19. Oden's suggestion ("Baʿal Šamēm" 462) that the three gods mentioned here are actually one and the same can hardly be taken seriously. See pp. 52-53.

20. See Oden, ibid. 459.

21. Teixidor, *The Pagan God: Popular Religion in the Greco-Roman Near East* (Princeton: Princeton University Press, 1977) 26.

22. M. Lidzbarski, *Ephemeris für semitische Epigraphik* (3 vols.; Giessen: Rickers-Töpelmann, 1902-1915) 1.248 n.1.

23. Oden ("Ba$^{c}$al Šamēm" 466) tries to argue that Ba$^{c}$al-Šamem is not distinct from Ba$^{c}$al-Ḥamon in this inscription. But see pp. 52-53.

24. Février, "Paralipomena punica" 13-22. For the present writer's position on the identification of *b$^{c}$l mgnm* see pp. 63-64.

25. Lambert, "Historical Development" 194.

26. Gsell 4.233-34; similarly Albright, *Archaeology* 82.

27. See p. 7.

28. See, for example, G. E. Mendenhall, "The Ancient in the Modern --and Vice Versa," *Michigan Oriental Studies in Honor of George G. Cameron* (ed. L. L. Orlin; Ann Arbor, Mich.: University of Michigan Press, 1976) 235-37.

29. See Cross, *Canaanite Myth* 33; Oden, "The Persistence of Canaanite Religion," *BA* 39 (1976) 34-36.

30. *AgAp* 1.113.

31. See p. 41.

32. Diodorus Siculus, *Bibliothēkē* 20.14.

33. The identification of Kronos with Ba$^{c}$al-Hamon is securely established. See W. Röllig, *WdM* 271. Note also *KAI* 176, which begins: *KRONŌI THENNEITH PHENĒ BAL*; the god here named Kronos can be none other than Ba$^{c}$al-Ḥamon, who is named at the beginning of innumerable Punic stelas along with Tanit (here *THENNEITH*).

34. "Les papyrus araméens d'Hermoupolis et les cultes syro-phéniciens," *Bib* 48 (1967) 566.

35. It is important to recognize that our knowledge of the pantheon of Tyre--whether "official" or "popular"--during this period is all but non-existent. There are no inscriptions from Tyre dating to this period. Later inscriptions mention Melqart as the "Lord of Tyre" (*KAI* 47 [from Malta], second century B.C.), $^{c}$Aštart (*KAI* 17 [from Tyre], second century B.C.), and Ba$^{c}$al-Šamem (*KAI* 18 [from Umm el-$^{c}$Awāmīd, near Tyre], 132 B.C.). Though none of these sources is an official inscription, and although they derive from a much later period, they are evidence at least that three of the deities named in E/B were worshipped in the vicinity of Tyre in the pre-Christian period. Given the gap in our knowledge of Tyrian religion during this period, it is gratuitous to assume, as Milik does (see n.34), that other deities mentioned in the foreign god section of this god-list were not Tyrian.

36. "Paralipomena punica" 16.

37. Written $^{d}$ŠUḪ.ŠEŠ; see M. Streck, *Assurbanipal und die letzten assyrischen Könige bis zum Untergange Niniveh's* (3 vols.; Vorderasiatische Bibliothek 7; Leipzig: J. C. Hinrichs, 1916) 2.52.30.

38. See Chapter III.

39. *WdM* 55.

40. R. Labat, *CAH*[3] 2/2, 401.

41. Ibid.

42. Borger, *Inschriften* 84.42 (AN.GAL $^{d}$*Šar-rat* BÀD.AN$^{ki}$).

43. Frankena, *Tākultu* 113: "AN-GAL ( = $^{d}$Sataran), the chief god of Dēr"; E. F. Weidner, "Die Feldzüge Šamši-Adads V. gegen Babylonien," *AfO* 9 (1933-1934) 99: "Šarrat of Dêru (Dêrîtum) was the spouse of the god KA.DI [ = Sataran] and the chief goddess of Dêru."

44. Weidner, ibid.

45. Ibid. 98-99.

46. Written $^{d}$*A-tar-sa-ma-a-in*; see Borger, *Inschriften* 53.10 (on the writing of $^{c}$Ataršamain, see Appendix I, n. 19).

47. There are several reasons for taking $^{c}$Ataršamin as the supreme deity of the Arabs. (1) Assurbanipal addresses a prayer to this goddess (see M. Cogan, *Imperialism and Religion: Assyria, Judah and Israel in the Eighth and Seventh Centuries B.C.E.* [SBLMS 19; Missoula, Mont.: Scholars Press, 1974] 15-20). In it he thanks her for handing over Hazail to his grandfather, Sennacherib, and deserting the people of Arabia for Assyria. The fact that Assurbanipal dedicates a prayer to $^{c}$Ataršamain and that she wields so much power in the affairs of Arabia points to her very high status among the gods of Arabia. (2) When Esarhaddon returns the gods of Arabia to Hazail, he says that Hazail "besought me to return his *gods*." Then six Arabian deities are listed, beginning with $^{c}$Ataršamain (Borger, *Inschriften* 53.6-12). Later, when Assurbanipal refers to this event in the above-mentioned prayer, he simply says that Hazail "appealed to him [Esarhaddon] concerning the return of his *goddess*" (Cogan, 17.11). This seems to imply that $^{c}$Ataršamain was the most important of these deities, since Hazail was anxious to secure the return of her image in particular. Note that later still, when Assurbanipal returns these gods to Hazail's son, Uate', again only $^{c}$Ataršamain is mentioned (Cogan, 36). (3) Most importantly, Assurbanipal calls the Arabs the "people of $^{c}$Ataršamain" ($^{lú}$*a'-lu šá* $^{d}$*A-tar-sa-ma-a-a-in* [Streck, 2.72.112,124]; cf. Hebrew $^{c}$*m yhwh*, "the people of Yahweh," as a designation of the Israelites [see *THAT* 2.302-5]); see W. Caskel, "Die alten semitischen Gottheiten in Arabien," *Le Antiche Divinità Semitiche* 100, and W. J. Dumbrell, "The Tell el-Maskhuta Bowls and the 'Kingdom' of Qedar in the Persian Period," *BASOR* 203 (1971) 44 n. 48. I am grateful to J. S. Kselman for bringing this last reference to my attention.

48. Wiseman, *Chronicles of Chaldean Kings (626-556 B.C.) in the British Museum* (London: British Museum Publications, 1956) 32.

49. Cogan, *Imperialism and Religion* 25.

50. Ibid. 23.

51. Ibid. 24.

52. *Yahweh and the Gods of Canaan: A Historical Analysis of Two Contrasting Faiths* (Garden City, N.Y.: Doubleday, 1968) 227.

53. McCarthy, *Treaty and Covenant* 73 (rev. ed. 114).

54. *Pagan God* 32.

55. There has been some hesitation as to the correct reading of this divine name even after Borger's latest publication of the text based on a reexamination of the photographs (*Inschriften* 107); but Borger has defended the reading $^{d}$*A-na-ti* in a later article, "Anath-Bethel," *VT* 7 (1957) 102-4.

56. "Ba$^{c}$al Šamēm" 463. Albright also explicitly calls Ba$^{c}$al-Šamem "the head of the Tyrian pantheon" in *Yahweh* 239.

57. "Vassal-Treaties" 130.

58. "Die Götteranrufung in den keilschriftlichen Staatsverträgen," *Or* 45 (1976) 127.

59. E.g., Jer 48:13.

60. F. M. Cross, Jr., and D. N. Freedman, *Early Hebrew Orthography: A Study of the Epigraphic Evidence* (AOS 36; New Haven, Conn.: American Oriental Society, 1952) 14, 19; Z. S. Harris, *A Grammar of the Phoenician Language* (AOS 8; New Haven, Conn.: American Oriental Society, 1936) 37; J. Friedrich and W. Röllig, *Phönizisch-Punische Grammatik* (rev. ed.; AnOr 46; Rome: Pontificium Institutum Biblicum, 1970) 32.

61. *CAD* "B" 282a; *AHW* 132b.

62. Albright, *Archaeology* 170.

63. See Cross and Freedman, *Orthography* 27-29; Fitzmyer, *Sefîre* 146.

64. Ibid.

65. For *Bētylos* as a variant of *Baitylos*, see Friedrich and Röllig, §241. By the time of Eusebius *ai* and *e* in Greek were homophonous (see BDF §25); this might mean that the first syllable of this word was pronounced /be/ or /bē/ rather than /bay/.

66. *Pr. Ev.* 1.10.16.

67. See A. Cowley, *Aramaic Papyri of the Fifth Century B.C.* (Osnabrück: Zeller, 1967 [reprint of the 1923 edition]) 2.6,10; 12.9; 18.4,5; 22.6; 42.8; 55.7.

68. M. D. Coogan, *West Semitic Personal Names in the Murašû Documents* (HSM 7; Missoula, Mont.: Scholars Press, 1976) 46-47.

69. Ibid. 45. Although the documents Coogan is concerned with date from the fifth century B.C., there is evidence that "the use of DINGIR and DINGIR.MEŠ as free variants is not restricted to the Neo-Babylonian period . . ." (p. 58 n. 19) and thus could well have been in practice in the seventh century B.C., when E/B was written.

70. Fitzmyer, *Sefîre* 90; McCarthy, *Treaty and Covenant* 73 n. 17 (rev. ed. 114 n. 25).

71. *Archaeology* 170.

72. That Bethel's veneration by the Arameans does not necessarily mean that he was an Aramaic deity may be supported by further evidence. There are two other possible instances of the worship of a Phoenician (Tyrian?) deity by Arameans in the first half of the first millennium B.C. Two royal stelas from the general vicinity of Aleppo, dating from the ninth century B.C., honor gods that may be Phoenician. The Zakkur stela (*KAI* 202), found 45 km. southwest of Aleppo, mentions Ba$^{c}$al-Šamem (in the Aramaized form *b$^{c}$lšmyn*) a number of times. It is possible that he is not a Phoenician deity here, although H. J. Katzenstein believes he is (*The History of Tyre: From the Beginning of the Second Millenium* [*sic*] *B.C.E. until the Fall of the Neo-Babylonian Empire in 538 B.C.E.* [Jerusalem: The Schocken Institute for Jewish Research, 1973] 200). Another ninth-century stela (*KAI* 201), found 7 km. north of Aleppo, is dedicated by the Aramean king Bir-Hadad to the god Melqart. Katzenstein identifies the god here as the Tyrian Melqart (ibid. 138). Albright has expressed reservations about Melqart's being a specifically Tyrian deity (Katzenstein, 139 n. 57; Albright, *Archaeology* 196 n. 29). But aside from this inscription the divine name is not attested except at Tyre and in her colonies. The name itself is Phoenician, not Aramaic. The element *qrt* ("city") is attested in Phoenician and Punic (*DISO* 267) but not in Aramaic; the corresponding Aramaic word is *qry*ʾ/*h*, construct *qryt* (*DISO* 266), not *qrt* (*qart*-).

73. J. P. Hyatt ("The Deity Bethel and the Old Testament," *JAOS* 59 [1939] 81-98) has argued that the divine name is attested at Ugarit. He cites *btʾil* (*UT* 31[14].1) and the proper name *n$^{c}$btʾil* (*UT* 101[10].16). But M. Pope (*ʾEl in the Ugaritic Texts* [VTSup 2; Leiden: Brill, 1955] 59-60) has convincingly demonstrated the weaknesses of Hyatt's hypothesis; for example, the name *n$^{c}$btʾil* is most likely a misreading (p. 59). Pope concludes: "There is thus not a single instance in the Ugaritic texts where the words *bt ʾil* may, with any degree of probability, refer to the god Bethel" (p. 60). F. (Quest-)Gröndahl (*Die Personennamen der Texte aus Ugarit* [Studia Pohl 1; Rome: Päpstliches Bibelinstitut, 1967] 118) lists a proper name *Be-ti/bi-ti*$_{4}$-DINGIR$^{lim}$/*ʾi-lu*, which she compares to Bethel. But the first element of the name could as well be read *Bî-ti/bi-ti*$_{4}$, i.e., **bitti*, "daughter of (X)," as in the name $^{f}$*Bitta-rapiʾ* (p. 180).

74. *Pr. Ev.* 1.10.16.

75. Zadok, "Phoenicians, Philistines, and Moabites in Mesopotamia," *BASOR* 230 (1978) 61.

76. See F. L. Benz, *Personal Names in the Phoenician and Punic Inscriptions: A Catalog, Grammatical Study and Glossary of Elements* (Studia Pohl 8; Rome: Biblical Institute, 1972) 382.

77. Cowley, *Aramaic Papyri* 279, 304.

78. Cf. *mlk$^{c}$štrt*, Milk-$^{c}$Aštart" (*KAI* 19.2-3; 71.2; 119.1); *ṣdtnt*, "Ṣid-Tanit" (*CIS* 1.247-49); *ṣdmlqrt*, "Ṣid-Melqart" (*CIS* 1.156); *ʾšmn$^{c}$štrt*, "ʾEšmun-$^{c}$Aštart" (*CIS* 1.245.3-4); *ʾšmnmlqrt*, "ʾEšmun-Melqart" (*CIS* 1.16), etc.

79. *Pr. Ev.* 1.10.16.

80. E. Kraeling, *The Brooklyn Museum Aramaic Papyri: New Documents of the Fifth Century B.C. from the Jewish Colony at Elephantine* (New Haven, Conn.: Yale University Press, 1953) 89 and Cross, *Canaanite Myth* 47 n. 14.

81. Hyatt, "Bethel" 98.

82. Kraeling, *Aramaic Papyri* 89. See also Dhorme and Dussaud, 361.

83. B. Porten, *Archives from Elephantine: The Life of an Ancient Jewish Military Colony* (Berkeley/Los Angeles: University of California Press, 1968) 135.

84. Cross, *Canaanite Myth* 30 n. 102.

85. Cf. *šm yhwh*, "the Name of Yahweh"; *kbwd yhwh*, "the Glory of Yahweh"; *rwḥ yhwh*, "the Spirit of Yahweh"; *pny yhwh*, "the Face of Yahweh"; etc.

86. Albright, *Archaeology* 174.

87. The god Bethel was himself further hypostatized at a later date to *šm-byt*ʾ*l*, "the Name of Bethel." See Milik, 567.

88. I.e., the divine determinative (DINGIR) is never used with these temple names; see Hyatt "Bethel", 92-93.

89. R. Dussaud, *Les religions des hittites et des hourrites, des phéniciens et des syriens* 361.

90. Something similar appears to have taken place in the case of *ml*ʾ*k yhwh* ("the angel/messenger of Yahweh") in the OT. "It seems inescapable that the *mal*ʾ*ak yhwh* is in some cases an expression for God himself. . . . *Mal*ʾ*ak* really meant a 'presence' or re-present-ation of God, earlier than it meant re-present-ative, or simultaneously" (R. North, "Separated Spiritual Substances in the Old Testament," *CBQ* 29 [1967] 438).

91. Hyatt, "Bethel" 91.

92. RS 20.24 and RS 26.142; see J. Nougayrol, *Ugaritica V* 42-64. Here the gods are arranged in a hierarchical sequence (p. 42) "in two 'rows,' one led by the supreme god of the place, the other by his consort" (p. 44). Nougayrol is puzzled by the fact that the list is translated into Akkadian; he suggests, "One could suppose . . . that we are dealing with *un document de la pratique* rather than a theological text . . ." (p. 43). Perhaps the translation of the document can be explained by regarding it as a god-list connected with the treaty tradition at Ugarit. One should note that immediately after the mention of the high god *b*ᶜ*l*, the storm-god, there is listed a series of "Baᶜals" (pp. 44-45), just as a series of "Tešubs" (the Hurrian storm-god) is listed immediately after Tešub (ᵈIM/ᵈU) in the Hittite treaties. See also nn. 373 and 463 below for more parallels between this list and the Hittite treaty god-lists.

93. The Akkadian translation of the god-list has DINGIR *a-bi*, which could mean "god of (the) father" or, according to Laroche, "god-father"

(*Ugaritica V* 46). But the Hurrian equivalent, also found in god-lists at Ugarit, clearly does not mean "god of the father." The phrase occurs in the singular, *ʾin ʾatnd (eni attanni-da)*, "to god the father" or "to the divine father (?)" (RS 24.254.2; *Ugaritica V* 507), as well as in the plural, *ʾinšt ʾatn[št]nm (enna-šta attanna-šta-ma)*, "and to gods the fathers" or "and to the divine fathers (?)" (RS 24.261.12; *Ugaritica V* 499). One may draw two conclusions from these observations: (1) Ugaritic *ʾil ʾib* is not to be interpreted as "god of the/my father" as if it referred to some patron deity; (2) the fact that the phrase also occurs (in Hurrian) in the plural indicates that in no case is *ʾil ʾib* to be understood as a divine proper name, that is, a god who would rank above 'El in the Ugaritic pantheon. Compare Hittite *attaš siuneš* ("gods-father" or "dieux paternels"), which according to Laroche are the divine shades ("Recherches" 72).

94. *Pr. Ev.* 1.10.16.

95. *Pr. Ev.* 1.10.31.

96. See Kraeling, *Aramaic Papyri* 89.

97. See O. Eissfeldt, "Der Gott Bethel," *Kleine Schriften* (ed. R. Sellham and F. Maass; 5 vols.; Tübingen: Mohr, 1962) 1.224. Albright has maintained that the first element of the name is a common noun, something like "Sign [of Bethel]" or "Will [of Bethel]" (*Archaeology* 174), not the proper name ᶜAnat. He has been followed on this point by several scholars (Milik, "Les papyrus araméns" 566-67; Teixidor, *Pagan God* 31). Albright's view is clearly based on grammatical considerations, since at least in Hebrew a proper noun cannot be in the construct state (see, for example, P. P. Joüon, *Grammaire de l'hébreu biblique* [Rome: Institut Biblique Pontifical, 1923] §§131n-o, 137b); hence *A-na-ti* here would be a common noun.

But this grammatical law may not be so absolute as Albright seems to think. Recently M. Liverani ("Preistoria" 331-34) has challenged the contention that the proper name in *yhwh ṣbʾwt* cannot be in construct (Cross, *Canaanite Myth* 65). One finds not only ᵈIM KARAŠ ("Tešub of the army") in the Hittite treaties but also Ugaritic *ršp ṣbʾi* (*UT* 2004.15), which may mean "Rešep of the army." In any case, there is quite enough evidence available at present from which to argue convincingly that proper names do occur in construct in West Semitic. One should note, for example, Phoenician *ršp ḥṣ* (*KAI* 32.3-4), "Rešep of the arrow," or *ᶜttrt šd* (*UT* 2004.10), clearly "ᶜAštart of the field" (cf. ᵈ*Ištar ṣēri*, "Ištar of the field"). There is even more evidence from Ugaritic and Phoenician proper names having the pattern "DN of GN." Note *ʾatrt ṣrm* (*CTA* 14[KRT].4.192, 201), "Ašerah of Tyre"; *ršp bbt* (612.A.11), "Rešep of BBT" (cf. *ršp bbth* [607.31] with the *he locale*); and Phoenician *ᶜštrt kt* (*KAI* 37 A.5) "ᶜAštart of Kition." Other examples could be adduced. But the ones mentioned here seem sufficient to establish the point that in the name ᶜAnat-Bethel the element ᶜAnat is a proper name in construct with Bethel: "ᶜAnat (consort) of Bethel." Similarly the name *ᶜntyhw* Cowley, *Aramaic Papyri* 44.3), " "ᶜAnat-Yahu," undoubtedly means "ᶜAnat (consort) of Yahu," as Cowley suggested over fifty years ago (p. 148).

98. Cross, "ʾēl," *TDOT* 1.253.

99. Landsberger, *Sam'al: Studien zur Entdeckung der Ruinenstätte Karatepe* (Ankara: Türkische historische Gesellschaft, 1948) 47 n. 117.

100. Diodorus Siculus, *Bibliothēkē* 20.14.4-7.

101. *Canaanite Myth* 26.

102. *Pr. Ev.* 1.10.16,20,29,44.

103. Cross argues that the element *ḥmn* in *b*ᶜ*l ḥmn* is equivalent to Ugaritic *ḫmn* (Akkadian KUR *Ḫa-ma-ni*), the name of the Amanus mountain range in extreme northwest Syria (*Canaanite Myth* 26-28). He finds one piece of evidence connecting this place name with 'El in a Hurrian hymn to this god (see E. Laroche, *Ugaritica V* 510-16) containing the phrase *'il pbnḫwm / ḫmn*, perhaps "'El, the one of the mountain / Ḫamān"; Cross (28 n. 85) admits, however, that the syntax is not altogether clear. Against this interpretation see J.J.M. Roberts, "El," *IDBSup* 256b and idem, "Davidic Origin" 336 and nn. 51-52. Also damaging to Cross's thesis is the fact that in another Hurrian hymn from Ugarit (*CTA* 166.50-51) the name of the god Nubadig is also immediately followed by the word *ḫmn* (*nbdg / ḫmn*).

104. "Baᶜal Šamēm."

105. See p. 39.

106. Oden's sources for the argument from divine epithets are: an Aramaic boundary inscription from Gözneh (*KAI* 259), fourth/fifth century B.C. (Oden 465); Aramaic inscriptions from Hatra (esp. *KAI* 244) first/second century A.D. (ibid. 467); Palmyrene inscriptions, also from this period (ibid. 468); and a citation from Isaac of Antioch, fifth/sixth century A.D. (ibid. 469). His sources for the argument from iconography are all from the Christian era.

107. Ibid. 465.

108. See W. J. Fulco, *The Canaanite God Rešep* (AOS Essay 8; New Haven, Conn.: American Oriental Society, 1976) 8.

109. See pp. 38-40 and Chapter IV n. 11.

110. Teixidor (*The Pantheon of Palmyra* [EPROER 79; Leiden: E. J. Brill, 1979] 26) argues that *qn'* does not mean "create" but rather "possess" in Aramaic. But in all likelihood the epithet here has been taken over mechanically from the Phoenician *qnh 'rṣ*.

111. See n. 81.

112. Teixidor, *Pagan God* 28, 54. Oden does not discuss this epithet, which militates against his view that Baᶜal Šamem is 'El.

113. See, for example, *WdM*, pl. 6 no. 8 (between pp. 312 and 313).

114. Oden, "Baᶜal Šamēm" 471.

115. For the text, see p. 41.

116. Oden, "Ba$^{c}$al Šamēm" 466.

117. Ibid.

118. For the text, see p. 41.

119. Ibid. 462.

120. Ibid. 461, 470.

121. Ibid. 470.

122. Ibid. 459, 461.

123. Ibid. 472.

124. *Pr. Ev.* 1.10.17 (my translation).

125. Cross, *Canaanite Myth* 7 n. 13.

126. Oden, "Ba$^{c}$al Šamēm" 464: "the 'syncrétisme solaire' of the Hellenistic and later eras meant that *all* chief deities were identified with the sun" (emphasis mine). See also H. Seyrig, "Le culte de Bêl et de Baal shamîm à Palmyre," *Antiquités syriennes* (Paris: 1934) 1.94. See also Teixidor, *Pagan God* 47.

127. *KAI* 2.68.

128. Oden, "Ba$^{c}$al Šamēm" 464.

129. Translation of H. L. Ginsberg, *ANET*$^{3}$ 153b.

130. Hillers, *Treaty Curses* 13.

131. See D. O. Edzard, *WdM* 84.

132. Rev IV.11.

133. Rev IV.12: $^{giš}$*tar-kul-la-ši-na li-is-su-ḫu*, "May they tear up their [i.e., the ships'] mooring poles."

134. *Pagan God* 32.

135. Rev IV.5.

136. For example, in the Erra epic "the Sibitti are seven wicked gods, without individual names, who act as a unit to the point that, more than once, the agreement of grammatical forms is in the singular instead of the plural" (L. Cagni, *The Poem of Erra* 18).

137. Albright, *Yahweh* 127.

138. Pp. 84-86.

139. See Tallqvist, *Akkadische Götterepitheta* 54.

140. $^{d}$U AN *ù* KI (W2 Rev 40); $^{d}$U *ša-me-e* (W3 IV.9?; F6 I.41; KBo IV 10 Obv 51); $^{d}$IM *ša-me-e*/AN$^{(e)}$ (RS 17.146 Rev 49; 227 Rev 51´; 237 Rev 11´; 340 Rev 15´).

141. *Sam'al* 47 n. 117.

142. See *KAI* 2.35, 214, 223.

143. Teixidor, *Pagan God* 28.

144. See, for example, Picard, *Hannibal* 29.

145. See n. 102 above.

146. *Pr. Ev.* 1.10.7.

147. See M. Höfner, *WdM* 429.

148. See Chapter II.

149. *Pr. Ev.* 1.10.29; Hesiod, *Theogony* 154-210.

150. See Xella, "Giuramento annibalico" 192-93.

151. Moscati, *Fenici* 536.

152. *WdM* 480.

153. E. Riess, "Ammon," PW 1 col. 1855.

154. In several recent publications Cross has stated that "the regular order of treaty witnesses" is "patron gods, high gods, old gods . . ." (*Canaanite Myth* 10 n. 32; see also "Olden Gods" 334 and "*'ēl*" 251). He is referring here specifically to Sf1. But this analysis of the order of divine witnesses is incorrect for Sf1 as well as for all other ancient Near Eastern treaties. All Mesopotamian treaties list the supreme god (Marduk in Babylonia, Assur in Assyria) first. All Hittite treaties begin with the sun-god of heaven and/or the sun-goddess of Arinna; although the sun-goddess may conceivably be called a "patron deity" and although she is listed with the king's divine helpers in war, the sun-god never is (Gurney, *Hittite Religion* 6). Since the first half of the Sf1 god-list is incontrovertibly based on Mesopotamian treaty tradition, it would be strange if the first deities listed ([DN] *wmlš*) were not the supreme deities --that is, of KTK.

There is more than enough evidence to show that "patron" or "personal" deities of kings were listed in a specific slot within god-lists, *after* the supreme deities. Recently Kestemont has noted that in Hittite treaties the place for such deities is just after the "principal Hittite deities" ("Panthéon" 148, 157; see Appendix I, Chart 14). In other official texts from Hatti the same pattern is followed. On the relief at Yazilikaya, which dates from the time of Tudḫaliyas IV, we find the order Tešub, Hepat (the supreme consort pair of the Hurrian pantheon) followed by Šarruma, the personal god of Tudḫaliyas (see Gurney, *Hittite Religion* 19). In the annals of Mursilis II, his personal god Mezulla (see Laroche, "Recherches" 30) is always mentioned just after the supreme gods, the sun-goddess of Arinna and her consort the storm-god (Goetze, *Die Annalen des Muršiliš*

22, 32, 42, 44, 50, etc.). In the "imperial pantheon" list from the Ur III Dynasty (Appendix I, Chart 2), Il-aba, the "patron deity of the Sargonic dynasty" (J.J.M. Roberts, *The Earliest Semitic Pantheon: A Study of the Semitic Deities Attested in Mesopotamia before Ur III* [Baltimore: Johns Hopkins University Press, 1972] 149), appears just after the highest gods--Ištar Annunitum, Anu, and Enlil. The same pattern is followed (*pace* Cross) in the inscriptions from Šam'al (*KAI* 24, 214, 215). Here Rakib-'el, who is specifically designated *b$^{c}$l bt*, "lord of the house [i.e., dynasty]" (*KAI* 24.16; 215.22), always comes after the highest gods, Hadad and 'El (see n. 11). Hence the order of divine witnesses in treaties is supreme gods,(patron gods), high gods--*not* patron gods, high gods.

155. "Paralipomena punica" 17.

156. Ibid. 16.

157. See Chapter I.

158. A/M Rev VI.7-12; Sfl A.8-9; VTE(A,B,C) (i.e., Assur and Ninlil [16,19; 25,29; 414,417]; see Edzard, *WdM* 44). Cf. the "pantheon list(s)" from Ugarit, which may derive from the treaty tradition (see n. 92 above): the order is *'il . . . 'aṯrt*, "'El . . . 'Ašerah."

159. On the identification of Tanit with Hera or Juno, see Cross, *Canaanite Myth* 29; Gsell 4.255-58; of $^{c}$Aštart with Hera or Juno, see Gsell 4.255-58 and Fitzmyer, "The Phoenician Inscription from Pyrgi" 288.

160. Moscati (*Fenici* 536) says "Hera [in this text] is without doubt Tanit."

161. On the identification of Athena with $^{c}$Anat, see *KAI* 42.1; R. du Mesnil du Buisson, *Nouvelles études sur les dieux et les mythes de Canaan* (EPROER 33; Leiden: Brill, 1973) 48-55.

162. *Pr. Ev.* 1.10.18,32.

163. See, for example, A. Caquot et al., eds., *Textes ougaritiques: Tome I, Mythes et légendes* (LAPO 7; Paris: Editions du Cerf, 1974) 85.

164. Cf. the Ugaritic goddess Pidray (*pdry*), who was both daughter and consort to Ba$^{c}$al (see Albright, *Yahweh* 125 n. 38).

165. Oden has claimed that $^{c}$Anat and $^{c}$Aštart were fused into a conflate deity "$^{c}$Anat-and-$^{c}$Aštart" as early as the period of Ugarit. He cites 607.20 ( = RS 24.244.20), in which *$^{c}$nt w$^{c}$ṯtrt* occurs without the word divider ("Persistence" 34). But the lack of the word-divider here is not so significant as Oden seems to think. For example in the same text we read *kṯr.wḫss* (46),"Kôṯar-and-Ḫasis"; it is absolutely clear from the epic literature that this is a single god, yet the word divider occurs here. Moreover, in the other texts (601.1.22; 608.14) where the names of $^{c}$Anat and $^{c}$Aštart are juxtaposed in this sequence the word divider does appear: *$^{c}$nt.w$^{c}$ṯtrt*.

166. *CTA* 14[KRT].6.291-93; cf. *CTA* 14[KRT].3.145-46.

167. Note that here, as in *$^{c}$nt.w$^{c}$ṯtrt*, *$^{c}$nt* is the "A-word" and *$^{c}$ṯtrt* forms the "B-word."

168. *s̀šw.*$^{c}$*ttrt.ws̀šw.*$^{c}$*[nt] (UT* 2158.1.6*);* $^{c}$*ṯtrt.w*$^{c}$*nt* (601.1.9); [$^{c}$*ṯ*]*trt w*$^{c}$*nt* (601.2.1); *l.*$^{c}$*ṯtrt.l.*$^{c}$*nt (UT* 2008.1.7-8*)*. See also n. 165 above.

169. See W. Helck, *WdM* 333.

170. J. B. Pritchard, *Recovering Sarepta, A Phoenician City: Excavations at Sarafand, Lebanon, 1969-1974, by the University Museum of the University of Pennsylvania* (Princeton: Princeton University Press, 1978) 104-5.

171. S. Moscati, *The World of the Phoenicians* (tr. A. Hamilton; London: Weidenfeld and Nicolson, 1968) 139; D. Harden, *The Phoenicians* (London: Thames and Hudson, 1963) 87.

172. Moscati, *Phoenicians* ibid.; Harden, *Phoenicians* ibid.; Gsell 4.264.

173. Gsell 4.264. However, the element mlkt could just as easily refer to $^{c}$Aštart; see Benz, *Personal Names* 346.

174. Benz, *Personal Names* 312-13; in this case, however, the element might refer to the Amanus as a sacred mountain (see Cross, *Canaanite Myth* 27 n. 77).

175. Benz, *Personal Names* 266-67. It is theoretically possible that the element *b*$^{c}$*l*, which is extremely frequent in Punic names, may sometimes refer to Ba$^{c}$al-Ḥamon; but Benz is probably correct when he associates the element solely with Ba$^{c}$al-Hadad (p. 228).

176. Note that similarly *ʾaṯrt* (ʾAšerah), the name of the chief goddess at Ugarit, rarely occurs in Ugaritic proper names according to J.C. de Moor ("*ʾ*$^{a}$*shērāh*," *TDOT* 1.440).

177. It is true, however, that we have little inscriptional evidence from Tyre and therefore not many personal names from that area. See n. 35.

178. See n. 77 above.

179. Note the remarks of Moscati, *Fenici* 533:

> It suffices to consider the cases of Tanit and Baal Hammon, which make clear beyond a possible doubt the non-correspondence between the onomasticon and the pantheon: whereas these two deities--as we know from the evidence of inscriptions--are the most popular at Carthage (or at least in the *tofet* of the city), the first appears in about four proper names, and the second does not appear at all, so that it is necessary to seek him in the generic Baal.

180. See G. Garbini, "Le iscrizioni puniche," *Missione archeologica italiana a Malta: Rapporto preliminare della campagna 1963* (ed. V. Bonello et al.; Rome: Centro di studi semitici, Instituto di studi del Vicino Oriente; Università di Roma, 1964) 83-89 and pl. 26.

181. See G. Levi della Vida in *RSO* 39 (1964) 318 (review of *Missione archeologica* [n. 180]).

182. Levi della Vida, ibid.; S. Moscati, "Astarte in Italia," *Rivista cultura classica e medioevale* 7 (1965) 757.

183. *Canaanite Myth* 30.

184. *Pace* Février, "Paralipomena punica" 17.

185. Oden, *De Syria Dea* 92.

186. The seventh-century B.C. inscription recently unearthed at Phoenician Sarepta, dedicated "to Tanit-ᶜAštart" (see n. 170), should perhaps also be understood to indicate the close association--rather than identification--of the two goddesses.

187. *Yahweh* 130. See also P. Berger, "La trinité carthaginoise: Mémoire sur un bandeau trouvé dans les environs de Batna et conservé au musée de Constantine," *Gazette archéologique* 5 (1879) 224.

188. Albright, *Yahweh* 130.

189. However, it is true that ᶜAštart also bore the epithet *caelestis/ourania*; see Gsell 4.262.

190. See Porten, *Archives from Elephantine* 165.

191. Ibid.

192. See Hvidberg-Hansen's dissertation, *La Déesse TNT: Une étude sur la religion canaanéo-punique* (Habilitationsschrift der Universität Kopenhagen; Kopenhagen, 1979). I am grateful to J. S. Kselman for this reference.

193. *Canaanite Myth* 31-35.

194. See p. 40.

195. *Canaanite Myth* 31.

196. Ibid. 33.

197. Cf. Pope, *WdM* 246, and de Moor, "'*ᵃšērāh*" 438. Again, once one realizes that proper names do occur in construct in Ugaritic (see n. 97 above) there is no reason that *'aṯrt ym* could not simply mean "'Ašerah of the sea." Cf. her consort's epithet, *'il.mbk.nhrm* (607.3, etc.), "'El of the confluence of the rivers," where the proper name *'il* is clearly in construct; or the Phoenician *ṣdn ym* (*KAI* 15), "Sidon of the sea" (cf. *KAI* 14.16: *ṣdn 'rṣ ym*).

198. *Canaanite Myth* 32.

199. Ibid.

200. Ibid. 33.

201. Cross believes that Tanit is mentioned as early as the Proto-Sinaitic inscriptions, which Albright has dated between 1525 and 1475 B.C. (*The Proto-Sinaitic Inscriptions and Their Decipherment* [HTS 22; Cambridge: Harvard University Press, 1969] 6), under the form *tnt* (*Canaanite Myth* 32). For a contrasting view, see Y. Yadin, "Symbols of Deities at Zinjirli, Carthage, and Hazor," *Near Eastern Archaeology in the Twentieth Century* [Nelson Glueck Festschrift] (ed. J. A. Sanders; Garden City, N.Y.: Doubleday, 1970) 229-30 n. 96.

202. *Proto-Sinaitic Inscriptions* 39.

203. *Canaanite Myth* 33.

204. If "Bašan" is the correct reading here, it would probably refer to the Biblical site, though not necessarily; the word itself seems to denote a "stoneless, fruitful plain," Arabic *baṯanat*.

205. Cf. for example the divine names/epithets in South Semitic such as *Dousarē* (*dū-š-Šara*), *dū-Ǵābat* in north and central Arabia (see M. Höfner, *WdM* 434-35, 438) and *ḏū-Maf*$^{c}$*alim*, *dū-bi-Raidān* in South Arabia (ibid. 514, 524). In the OT one finds *zh syny* < **z sny* < **ḏ sny*, "the one of Sinai" (Judg 5:5), an epithet of Yahweh.

206. *Canaanite Myth* 33.

207. Ibid.

208. Pp. 69-70.

209. *KAI* 39.3; 41.2; *RES* 1213.

210. Moscati, *Phoenicians* 141; Benedetto, "Divinità" 110; Walbank, *Historical Commentary* 47; Picard, *Hannibal* 10; R. Dussaud, "Astarté, Pontos et Ba$^{c}$al," *CRAIBL* (1947) 218; F. Vattioni, "Il dio Resheph," *AION* 15 (1965) 63.

211. Otten, "Lelvani" 120.

212. Février, "Paraliopmena punica" 20.

213. See Fulco, *Rešep* 50.

214. See Edzard, *Wdm* 109-10.

215. For this interpretation of *ḥs/ḥẓ* see Fulco, *Rešep* 51; Vattioni, "Reshep" 46; Liverani, "Preistoria" 333.

216. Appian, *Libykē* 127 (cited from Gsell 4.317 n. 4).

217. Valerius Maximus, *Memorabilia* 1.1.18.

218. Gsell 4.317.

219. Diodorus Siculus, *Bibliothēkē* 17.41.7-8.

220. Quintus Curtius, *Historiarum Alexandri Magni Macedonis* 4.3.22.

221. In *KAI* 214.2, however, *ršp* comes between *'l* and *rkb'l*. One should also compare *KAI* 24.15-16, where *rkb'l b*$^{c}$*l bt* is named after *b*$^{c}$*l ṣmd* ( = Hadad) and *b*$^{c}$*l ḥmn* ( = 'El).

222. Perhaps the best known of these triads was that of Ba$^{c}$al Šamem, $^{c}$Aglibol, and Malakbel; see, for example Höfner, *WdM* 420, 452.

223. Güterbock, "Hittite Religion" 85.

224. See, for example, Benedetto, "Divinità" 105-9; Walbank, *Historical Commentary* 46; A. Jirku, "Zweier-Gottheit und Dreier-Gottheit im altorientalischen Palästina-Syrien," *MUSJ* 45 (1969) 399-404.

225. *Pr. Ev.* 1.10.26.

226. "Paralipomena punica" 21.

227. *DISO* 142. The same word occurs in Hebrew as *māgēn*.

228. See my article, "$^{d}$LAMMA and Rešep" 466.

229. Fulco, *Rešep* 5-20.

230. "Solá Solé, "Miscelánea púnico-hispana I," *Sef* 16 (1956) 349.

231. *Altorientalische Forschungen* 443.

232. Cumont, "Gad," PW 7 cols. 433-35.

233. According to K. Ziegler ("Tyche," PW 14), *Tychē* is not only called a *daimōn* (col. 1646) but she is "particularly closely connected with the Agathos Daimon" (col. 1680); note that *Tychē* is often called *Agathē Tychē* (col. 1645)--cf. *Agathos Daimōn*.

234. Cumont, "Gad" col. 435.

235. Dussaud, "Astarté" 217 n. 2; Walbank, *Historical Commentary* 48.

236. "Gad" col. 434.

237. E.g., Picard, *Hannibal* 31; Moscati, *Fenici* 544; E. Groag, *Hannibal als Politiker* (Vienna: Seidel, 1929) 83 n. 2.

238. M. Höfner, *WdM* 438; G. Garbini, "Note di epigrafia punica--I," *RSO* 40 (1965) 213; *pace* G. Halff, "L'onomastique punique de Carthage, répertoire et commentaire," *Karthago* 12 (1965) 75.

239. *DISO* 47.

240. One finds *tychē* followed by a city name or a gentilic, with no indication of a difference of meaning: *Tychē Tarsou*, "Tyche of Tarsus" (O. Waser, "Tyche," *Ausführliches Lexikon* 5 col. 1371); but *Tychē Ephesiōn*, "Tyche of the Ephesians" (Waser, ibid.); *Tychē Antiochēon*, "Tyche of the Antiocheans" (ibid., col. 1365).

241. See Höfner, *WdM* 479.

242. See R. du Mesnil du Buisson, *Les tessères et les monnaies de Palmyre: Un art, une culture et une philosophie grecs dans les moules d'une cité et d'une religion sémitiques* (Paris: Editions E. de Boccard, 1962) 368.

243. Ibid. 215.

244. *Fenici* 536.

245. "Die Schutzgenien Lamassu und Schedu in der babylonisch-assyrischen Literatur," *Baghdader Mitteilungen* 3 (1967) 151.

246. Ibid.

247. See Gurney, *Hittite Religion* 8; Otten, "Yazilikaya" 29-30.

248. See H. Otten and W. von Soden, *Das akkadisch-hethitische Vokabular: KBo I 44 + KBo XIII 1* (StBoT 7; Wiesbaden: Harrassowitz, 1968) 27-32; *CAD* "L" 61a.

249. Benedetto, "Divinità" 114.

250. See p. 61.

251. Gsell 4.266.

252. "La plaquette en bronze d'Ibiza," *Sem* 4 (1951-1952) 28; "Inscripciones fenicias de la península ibérica," *Sef* 15 (1955) 50.

253. *KAI* 2.90.

254. Garbini, "Epigrafia punica" 213.

255. Ibid.

256. Guzzo Amadasi, *Le iscrizioni fenicie e puniche delle colonie in occidente* (StSem 28; Rome: Istituto di studi del Vicino Oriente; Università di Roma, 1967) 145.

257. "Astarté" 217 n. 2.

258. "Miscelánea" 350 n. 87.

259. Ibid.

260. Gsell 4.262.

261. Ibid. 261.

262. See A. García y Bellido, "Deidades semitas en la España," *Sef* 24 (1964) 237; idem, *Les religions orientales dans l'Espange romaine* (EPROER 5; Leiden: Brill, 1967) 140.

263. See A. Merlin, *Le sanctuaire de Baal et de Tanit près de Siagu* (Notes et documents publiés par la direction des antiquités et arts 4; Paris 1910) 42.

264. Ibid. 43.

265. On the identification of Ba$^{c}$al-Hamon with Saturn, see, for example, Gsell 4.283; Cross, *Canaanite Myth* 25.

266. So Merlin, *Le sanctuaire* 10.

267. See Gsell 4.273.

268. Oden, "Persistence" 34-35.

269. *Canaanite Myth* 35.

270. Cross, "The Origin and Early Evolution of the Alphabet," *Eretz-Israel* 8 (1967) 13*.

271. *KAI* 2.29.

272. *Personennamen* 154.

273. *CAD* "L" 23a.

274. Cf. also the Old Akkadian proper names *Ši-la-ba* and *Ištar-la-ba* (ibid. 25a).

275. RS 20.121.153 (*Ugaritica V* no. 119) and duplicates.

276. Merlin, *Le sanctuaire* 44-45.

277. Ibid. 47.

278. *Canaanite Myth* 35.

279. *Le sanctuaire* 43.

280. Ibid. 45.

281. *Canaanite Myth* 45.

282. See W. Helck, *WdM* 393.

283. G. Roeder, *Urkunden zur Religion des alten Ägypten* (Jena: Diederichs, 1915) 123; see W. Herrmann, "Aštart," *MIO* 15 (1969) 20.

284. *ANEP* 161 (no. 468).

285. R. du Mesnil du Buisson, *Etudes sur les dieux phéniciens hérités par l'empire romain* (EPROER 14; Leiden: Brill, 1970) 75. But no one else who has described the figure of $^{c}$Aštart on this seal has made such an observation; cf. W. F. Albright, "The Kyle Memorial Excavation at Bethel," *BASOR* 56 (1934) 8 (an excellent photograph of the seal impression may be found on p. 1 of the same article). One might therefore be inclined to dismiss du Buisson's description of the female figure as having a leonine head; yet the face of the goddess is oddly shaped, resembling (so it seems) more that of an animal than of a human, although it is not clearly leonine.

286. Hill, "Some Graeco-Roman Shrines," *JHS* 31 (1911) 58.

287. Ibid.

288. See G. F. Hill, *A Catalogue of the Greek Coins in the British Museum, vol. 27: Catalogue of the Greek Coins of Palestine (Galilee, Samaria, and Judaea)* (Bologna: Arnaldo Forni, 1965) lviii.

289. Gsell 4.271.

290. See P. Veyne, "'Tenir un buste': une intaille avec le Génie de Carthage, et le sardonyx de Livie à Vienne," *Cahiers de Byrsa* 8 (1958-59) 61-78.

291. See A. Fitzgerald, "The Mythological Background for the Presentation of Jerusalem as a Queen and False Worship as Adultery in the OT," *CBQ* 34 (1972) 413.

292. See, for example, the illustration in O. Waser, "Tyche" col. 1363.

293. For example, Waser, ibid. col. 1364, remarks: "the turret- or walled-crown . . . in particular designates the city-goddess. . . ." See also Baudissin, *Adonis und Esmun* 267; Gsell 4.259.

294. For example, at Caesarea-Arca (see G. F. Hill, *A Catalogue of the Greek Coins in the British Museum vol. 26: Catalogue of the Greek Coins of Phoenicia* [Bologna: Arnaldo Forni, 1965] 110) or Orthosia (ibid. 127).

295. Waser, "Tyche" col. 1365. For example, Beirut (Hill, *Phoenicia* 77, 90), Byblos (ibid. 99, 106), Sidon (ibid. 163, 168), Tripoli (ibid. 209, 213), Tyre (ibid. 254, 261).

296. Hill, "Shrines" 58.

297. Waser, "Tyche" col. 1371; Ziegler, "Tyche" cols. 1682, 1686.

298. See Baudissin, *Adonis und Esmun* 269 and pl. 6.

299. Ibid.

300. See *KAI* 175, where the name Ba$^{c}$al-Ḥamon is written in Greek *BAL AMOUN*.

301. See Röllig, *WdM* 271.

302. See Helck, *WdM* 331.

303. Baudissin, *Adonis und Esmun* 269; Gsell 4.259.

304. Ibid.

305. Baudissin, *Adonis und Esmun* 270.

306. J. Nougayrol, *PRU* 3.182 (RS 16.146 + 161).

307. *PRU* 3.182-83 n. 4.

308. Paul, "Jerusalem--a City of Gold," *IEJ* 17 (1967) 259-61.

309. Ibid. 261; see also idem, "Jerusalem of Gold--a Song and an Ancient Crown," *BAR* 3 (1977) 31.

310. "Jerusalem--a City" 260.

311. Ibid. 261-62; "Jerusalem of Gold" 39-40.

312. "The 'City of Gold' and the 'City of Silver'," *IEJ* 19 (1969) 178-80.

313. Ibid. 178-79.

314. Hoffner, "Jerusalem--a City" 261; "Jerusalem of Gold" 40.

315. Ibid. 40.

316. *Pr. Ev.* 1.10.27: *Melkathros, ho kai Hēraklēs*. Here some MSS read *Melkarthos*, which is closer to the Punic name.

317. Gsell 4.301.

318. Diodorus Siculus, *Bibliothēkē* 20.14.

319. See Benz, *Personal Names* 62 (*ʾmtmlqrt*), 75-81 (*bdmlqrt*), 104 (*grmlqrt*), 140-41 (*mlqrtḥlṣ*, etc.), 155-61 (*$^{c}$bdmlqrt*).

320. Gsell 4.303; Benedetto, "Divinità" 121; Walbank, *Historical Commentary* 47; Dussaud, "Astarté" 218; Moscati, *Phoenicians* 141 and *Fenici* 536.

321. Baudissin, *Adonis und Esmun* 286.

322. C. B. Avery, ed., *The New Century Classical Handbook* (New York: Appleton-Century-Crofts, 1962) 596.

323. Ibid.

324. H. T. Peck, ed., *Harper's Dictionary of Classical Literature and Antiquities* (New York: Cooper Square, 1962) 882a.

325. See p. 78 below and E. Lipiński, "La fête de l'ensevelissement et de la résurrection de Melqart," *Actes de la XVII$^{e}$ Rencontre assyriologique internationale: Bruxelles, 30 juin - 4 juillet 1969* (ed. A. Finet; Ham-sur-Heure: 1970) 37. It is not impossible that *mlk$^{c}$štrt* could mean "Milk of (the city) $^{c}$Aštart," especially if one compares the Ugaritic *mlk.$^{c}$ttrth* (607.41) and *mlk.b$^{c}$ttrt* (608.17), which clearly have this meaning (see Astour, "Serpent Charms" 21; *pace* M. Liverani, *OrAnt* 8 [1969] 340 [review of *Ugaritica V*], who terms Astour's position "perhaps simplistic"). Deities could retain very old epithets connecting them to cult centers even at much later times in areas far removed from the original cult center. Note for example *tnt blbnn*, "Tanit-in-Lebanon" at Carthage (*KAI* 81) or *$^{c}$štrt ḥr*, "$^{c}$Aštart-Ḥōr" (see F. M. Cross, "The Old Phoenician Inscription from Spain Dedicated to Hurrian Astarte," *HTR* 64 [1971] 190, 192), which could mean "$^{c}$Aštart of Syria" (cf. J. Teixidor, "A Note on the Phoenician

Inscription from Spain," *HTR* 68 [1975] 197, who believes that *ḥr* here is equivalent to Egyptian *ḫr* [as in *ᶜstr-ḫr*], "Phoenicia" or "Syria"; cf. Lucian, *De Syria Dea* 1, where *tēs Hērēs tēs assyriēs* refers to "Syrian Hera [ = ᶜAštart]"--see Fitzmyer, "Pyrgi" 288). Yet the interpretation "Milk(, consort) of ᶜAštart" is probably to be preferred.

326. Whereas in the Late Bronze Age (e.g., at Ugarit) these divine pairs are commonly linked by the copulative (*mt wšr, šhr wšlm*), in later periods the copulative is omitted: *ᶜṯtrt wᶜnt* > *ᶜtrᶜtʾ* = Atargatis.

327. On the identification of the "god of Hammon" with Milk-ᶜaštart, see M. Dunand and R. Duru, *Oumm el-ᶜAmed: une ville de l'époque hellénistique aux échelles de Tyr* (Paris: A. Maisonneuve, 1962) passim.

328. *Adonis und Esmun* 269.

329. Ibid. 264.

330. Dussaud, "Melqart," *Syria* 25 (1946-48) 228.

331. Ibid. 229. Although there is widespread agreement that *rṣp* here is *ršp*, this is by no means certain; if it is, it would certainly be a "very odd" writing of the name (see Fulco, *Rešep* 48; but note that the divine name *ʾšmn*, "ʾEšmun," once appears as *ʾzmn-* within a proper name [Benz, *Personal Names* 279]).

332. Vattioni, "Resheph" 65.

333. See Dussaud, "Melqart" 214-15.

334. Ibid.

335. Pausanias, *Description of Greece* 10.17.2.

336. Sardus is called *ho Makēridos* here.

337. Benedetto, "Divinità" 122; Dussaud, "Melqart" 214. As regards the form *Makēris* (a patronymic), cf. the name *Barmokaros* in the introductory section of H/P; according to Benz (*Personal Names* 348) the element *mokaros* is likewise a garbled form of Melqart.

338. F. Barreca, "Lo scavo del tempio," *Ricerche puniche ad Antas: Rapporto preliminare della Missione archeologica dell'Università di Roma e della Soprintendenza alle Antichità di Cagliari* (ed. E. Acquaro et al.; StSem 30; Istituto di studi del Vicino Oriente; Rome: Università de Roma, 1969) 33.

339. M. G. Guzzo Amadasi, "Note sul dio Sid," *Ricerche puniche ad Antas* 95.

340. Ibid. 101.

341. Athenaeus, *Deipnosophistai* 10.47.392d (Greek text in Baudissin, *Adonis und Esmun* 305).

342. Zenobius, *Paroemiographi* 5.56 (Greek text in Baudissin, *Adonis und Esmun* 305).

343. Baudissin, ibid. 307.

344. In a trilingual inscription from Sardinia (*KAI* 66) *l'šmn*, to 'Ešmun," is translated *Aesculapio* and *Asklēpiōi*. Note that the god is explicitly associated with healing here: "He ['Ešmun] heard his voice (and) *healed* him" (*šm[' q]l' rpy'*).

345. Thus if one must choose between Ṣid ( = Sardus) or 'Ešmun ( = Iolaos) as the god associated with Heracles/Melqart in H/P the choice must be 'Ešmun. However, there is at least a little evidence to suggest the possibility of an identification between the two Punic deities. Baudissin, for example, notes that "Sardos . . . corresponds to Iolaus, who was considered the founder of Sardinian culture . . ." (*Adonis und Esmun* 294), since both Sardus and Iolaos received the epithet *patēr* in Sardinia (p. 287); moreover, he points out that "according to Strabo (l[iber] V, 2, 7, C. 225) the mountain-dwellers of the island [i.e., Sardinia] were earlier called *Iolaeis* . . ." (p. 293). It is generally thought that *ṣd* was originally an epithet (Guzzo Amadasi, "Note sul dio Sid" 99), and it could mean "hunter" (Röllig, *WdM* 310). Damascius (in Photius, *Bibliotheca* 242.573H [cited from Baudissin, *Adonis und Esmun* 339]) designates *Ešmounos* ( ='Ešmun) as a hunter. Baudissin (p. 340) also observes that at Byblos Adonis was known as a hunter as was *Asklēpios*; 'Ešmun was certainly identified with the latter and perhaps also with the former (pp. 345-51). Thus Ṣid could have been another name for 'Ešmun as *ba^cal* (also originally an epithet) became a proper name for Hadad. But as yet there is not enough evidence to argue conclusively for the identification of *ṣd* and *'šmn*.

346. *Adonis und Esmun* 282-308.

347. Ibid. 302.

348. See pp. 76-77.

349. Cicero, *De Natura Deorum* 3.16.12 (Latin text in Baudissin, *Adonis und Esmun* 307).

350. Dussaud ("Melqart" 213) denies that Asteria here is equivalent to ^cAštart.

351. Among those who accept a consort relationship between Melqart and ^cAštart are R. du Mesnil du Buisson, *Nouvelles études* 58; Picard, *The Life and Death of Carthage* 46; Dussaud, "Melqart" 212; García y Bellido, *Religions orientales* 154; Lipiński, "Fête" 37; R. de Vaux, "Les prophètes de Baal sur le Mont Carmel,"*Bulletin du Musée de Beyrouth* 5 (1941) 7-20.

352. Lipiński, "Fête" 37.

353. Ibid.

354. Ibid. 38 n. 3.

355. Baudissin, *Adonis und Esmun* 308; see also Dussaud, "Astarté" 216.

356. "Divinità" 124.

357. Dussaud, "Astarté."

358. *Historical Commentary* 50.

359. See *The Oxford Classical Dictionary* (ed. M. Cary et al.; Oxford: Clarendon Press, 1949) 85b.

360. Ibid.

361. *Harper's Dictionary* 120b.

362. *Oxford Dictionary* 925b.

363. *Harper's Dictionary* 1608b.

364. Ibid.

365. *Oxford Dictionary* ibid.

366. *Harper's Dictionary* 1303b.

367. Ibid.

368. Ibid. 1304a.

369. "Astarté" 212.

370. Ibid. 212-13.

371. See G. and C. Picard, *La vie quotidienne à Carthage au temps d'Hannibal* (Paris: Hachette, 1958) 69 (cited from Vattioni, "Il dio Resheph" 62 n. 155).

372. Appendix I, Chart 14.

373. That *ym* in this Ugaritic list means "sea" and not the homographic "day" (as in Sf1) is clear from the Akkadian translation, *tâmtum* (A.AB.BA), "sea." See Nougayrol, *Ugaritica V* 45.

374. It is important to note that this inscription (the Arslan Tash amulet) is the only Phoenician text known to date to have close connections with the treaty tradition. Lines 10-11 explicitly assert, *krt.ln.ʾlt*, "(DN) has made a *treaty* with us"; see F. M. Cross and R. J. Saley, "Phoenician Incantations on a Plaque of the Seventh Century B.C. from Arslan Tash in Upper Syria," *BASOR* 197 (1970) 42-49. The "summary phrase" we are concerned with comes just after the name of several high gods (10: *ᶜlm* [ = ʾEl] and *ʾšr* [ = ʾAšerah]) and just before several divinized natural elements (13: *šmm.wʾrṣ.ᶜlm*, "Heaven and ancient Earth"); here the sequence high gods, summary phrase, natural elements is precisely the same as that found in the Hittite treaty god-lists. See most recently Z. Zevit, "A Phoenician Inscription and Biblical Covenant Theology," *IEJ* 27 (1977) 110-18.

375. *Die Phönizier* 2/2.468.

376. Ibid.

377. "Divinità" 124.

378. *Yahweh* 127 (emphasis mine).

379. See Albright, "Baal-Zephon," *Festschrift Alfred Bertholet zum 80. Geburtstag gewidmet von Kollegen und Freunden* (ed. W. Baumgartner; Tübingen: Mohr, 1950) 1.

380. Ibid. 11. See also O. Eissfeldt, *Baal Zaphon, Zeus Kasios und der Durchzug der Israeliten durchs Meer* (Halle: Niemeyer, 1932).

381. "Baal-Zephon" 8.

382. Ibid. 3.

383. Ibid. 11.

384. *Pr. Ev.* 1.10.14.

385. *Pr. Ev.* 1.10.20; see Albright, "Baal-Zephon" 12.

386. See p. 55.

387. *Yahweh* 128.

388. Cited from Harden, *Phoenicians* 174 (emphasis mine).

389. The text says that Hanno "set sail with 60 penteconters and 30,000 men and women, and provisions and other necessaries" (ibid.); whether or not these figures are exaggerated, the text clearly pictures Hanno's voyage as a major undertaking.

390. "Astarté" 218 n. 3. There is no doubt that at Ugarit Ba$^c$al (Hadad) and Mot (Death) were archenemies. It is possible that Adonis here could be Mot (see Pope, *WdM* 301-2).

391. P. D. Miller, Jr., *The Divine Warrior in Early Israel* (HMS 5; Cambridge: Harvard University Press, 1973) 24; on Ba$^c$al as a war-god, see ibid. 24-44.

392. Albright, *Archaeology* 195 n. 11. Albright thought the complete phrase was longer than these two words, but this has been disproven; see Miller, *Divine Warrior* 40.

393. Ibid. 41.

394. See n. 113 above.

395. Fulco, *Rešep* 2-22.

396. For a list of some of the authors who hold this position, see Oden, "Ba$^c$al Šamēm" 470 n. 71.

397. See E. Merkel, *WdM* 429.

398. *Yahweh* 127 (emphasis mine).

399. See, for example, Cross, *Canaanite Myth* 51.

400. In light of these arguments several scholars believe that Ares here is Hadad ( = Ba$^{c}$al-Šamem). If they are correct it is all the less probable that Poseidon in our list is Yamm ("Sea"). It would be very strange indeed for Hadad to be invoked in the same line with his archenemy Yamm--especially since Hadad is here conceived specifically in his role as the war-god, a role that brings to mind nothing so vividly as his victory over Yamm! A further reason for not understanding Poseidon as Yamm here is that "seas" may appear below in this god-list (see pp. 91-92).

401. "Astartê" 218-19.

402. *Historical Commentary* 50.

403. Pope, *WdM* 295.

404. Ibid. 296.

405. Ibid. 295.

406. Hence his equivalence in the Ugaritic "pantheon list(s)" with Ea, the Babylonian god of wisdom and incantations (see Nougayrol, *Ugaritica V* 45, 64; and Edzard, *WdM* 56).

407. *Pr. Ev.* 1.10.11.

408. *Religions orientales* 10.

409. *Harper's Dictionary* 789a.

410. *Religions orientales* 10.

411. "A Phoenician Treaty of Assarhaddon: Collation of K.3500," *RA* 26 (1929) 194.

412. See C. T. Lewis and C. Short, *A Latin Dictionary: Founded on Andrews' Edition of Freund's Latin Dictionary* (Oxford: Clarendon Press, 1879) 1102a.

413. See Langdon, "Assarhaddon": "The long vowel at the end is a mystery to me."

414. Cf. *aššurû/aššurê* (Assyrian dialect), "Assyrian."

415. Borger, *Assyrisch-babylonische Zeichenliste* 81; R. Labat, *Manuel d'épigraphie akkadienne (Signes, Syllabaire, Idéogrammes)* (Paris: Geuthner, 1976) 77[2].

416. *Fenici* 526.

417. C. Clemen, *Die phönizische Religion nach Philo von Byblos* (MVAG 42/3; Leipzig: J. C. Hinrichs, 1939) 50-51, denies that Zeus Meilichios here refers to Kušor, but rather--following Ewald and Baudissin--to his brother. It is true that while the preceding passage (10.11) speaks of "two brothers," only one is mentioned by name. But more recent treatments of Sakkunyaton's work accept the identification of Kušor with Zeus Meilichios. See Barr, "Philo of Byblos and His 'Phoenician History'," *BRJL* 57 (1974-

1975) 25; H. W. Attridge and R. A. Oden, Jr., *Philo of Byblos: The Phoenician History: Introduction, Critical Text, Translation, Notes* (CBQMS 9; The Catholic Biblical Association of America, 1981) 84.

418. *Pr. Ev.* 1.10.12.

419. LSJ 1093a.

420. See Riess, "Ammon" col. 1855.

421. See M. Avi-Yonah, "Syrian Gods at Ptolemais-Accho," *IEJ* 9 (1959) 6.

422. Eissfeldt, *Baal Zaphon* 7 n. 4; Clemen, *Phönizische Religion* 50; Moscati, *Fenici* 526.

423. Langdon, "Assarhaddon" 194.

424. Lewis and Short, *Latin Dictionary* 1120.

425. LSJ 1077b.

426. Moscati, *Fenici* 535; idem, *Phoenicians* 140; Février, "Paralipomena punica" 13; Picard, *Hannibal* 26; Groag, *Hannibal* 83; G. Egelhaaf, "Analekten zur Geschichte des zweiten punischen Krieges," *Historische Zeitschrift* 53 (1885) 458; Paton, *Polybius*, 421; E. S. Shuckburgh, *The Histories of Polybius* (New York: Macmillan, 1889) 515.

427. A.-T. Chroust, "International Treaties" 68.

428. *Historical Commentary* 50.

429. Ibid. 50, 52.

430. Ibid. 50.

431. See Chapter III n. 118.

432. See also Picard ("Carthage au temps d'Hannibal" 34 n. 2), who divides the god-list into the same three sections indicated in our analysis.

433. Note the juxtaposition of "abyss/deep" and "springs" in Prov 8:28 ($^{c}$*ynwt thwm*) and especially *thmwt* ("deeps") // *m*$^{c}$*ynwt* ("springs") in a cosmogonic context (Prov 8:24); cf. also *nkby-ym* ("the sources of the sea") // *thwm* in the same type of context (Job 38: 16). This juxtaposition also occurs in VTE 521: $^{d}$*E-a* LUGAL ZU.AB EN IDIM, "Ea, King of the *Deep*, Lord of *Springs*." D. R. Hillers ("Hear, O Heaven, and Give Ear, O Earth" 16) has argued--correctly, in our view--that the pairs of natural elements in the treaties are cosmogonic rather than theogonic.

434. L. R. Fisher, ed., *Ras Shamra Parallels: The Texts from Ugarit and the Hebrew Bible* (AnOr 49; Rome: Pontificium Institutum Biblicum, 1972) 1.367.

435. See Y. Avishur, "Studies of Stylistic Features Common to the Phoenician Inscriptions and the Bible," *UF* 8 (1976) 16.

436. Fisher, *Ras Shamra Parallels* 1.126-27; 2 (Rome: 1975) 399.

437. KBo V II 35 Obv II.12; KUB XXVI 39 Rev IV.24.

438. See n. 374.

439. Here *'rṣ wšmm*, "Earth and Heaven," occur as a pair; in this list one also finds *špš* ("Sun") and *yrḫ* ("Moon"), but not together. See Nougayrol, *Ugaritica V* 45.

440. Walbank, *Historical Commentary* 51.

441. Ibid.

442. Weinfeld, "*šbwᶜt*" 70 n. 175; idem, "Loyalty Oath" 397 n. 172.

443. Note also the sequence "heaven, earth, (and) sea" in Exod 20: 11; Hag 2:6; Neh 9:6.

444. L. Cagni, *L'Epopea di Erra* (StSem 34; Instituto di studi del Vicino Oriente; Rome: Università di Roma, 1969) 72-73; idem, *Poem* 32. In 1.170-71 [A.ME]Š ("waters") follows [AN$^{e}$ *u* KI$^{tim}$] (*Epopea* 76; *Poem* 34); and in 3 D.3-5 *tam-tam* ("sea") follows AN$^{e}$ . . . KI$^{tim}$ . . . *ma-tùm* ("land") (*Epopea* 100; *Poem* 48).

445. LSJ 1050b.

446. BAG$^2$ 475a.

447. LSJ ibid.

448. In the Hittite treaties this appears as A.AB.BA GAL (W3 Rev IV.44; W4 Rev I.10; RS 17.338 Obv 4; F1 D IV.18; KBo XII 31 Rev IV.14′) or Hittite *šalliš arunaš* (F4 B Rev IV.26; F5 IV.25?; F6 I.59 KUB XXIII 77a Obv 9; KBo IV 10 Rev 4; KUB XL 43 Rev 5; ABoT 56 II.30).

449. See p. 80.

450. For Hebrew *yam* with these connotations, see F. Zorell, *Lexicon Hebraicum et Aramaicum Veteris Testamenti* (Rome: Pontificium Institutum Biblicum, 1968) 313b. The word bears these nuances throughout Talmudic (see M. Jastrow, *A Dictionary of the Targumim, the Talmud Babli and Yerushalmi, and the Midrashic Literature* [New York: Jastrow, 1967] 597b) and modern Hebrew (see R. Alcalay, *The Complete Hebrew-English Dictionary* [rev. ed.; Bridgeport, Conn.: Prayer Book Press, 1974] 933. For Syriac *yammâ'* with the same meanings, see J. Payne Smith, *A Compendious Syriac Dictionary: Founded upon the Thesaurus Syriacus of R. Payne Smith, D.D.* (Oxford: Clarendon Press, 1903) 193a. Note too that the common Arabic word for "sea" (*baḥr*) also denotes a "large river" (see Wehr, *A Dictionary of Modern Written Arabic* 426).

451. BAG$^2$ ibid.; cf. Matt 4:18; Mark 1:16; John 21:1.

452. *JW* 3.506.

453. *Life* 96; 165; 304; *Ant* 14.450.

454. *DISO* 107.

455. See Fischer, *Ras Shamra Parallels* 1.203.

456. Hab 3:15; Pss 89:10-11; 93:4; etc.

457. E.g., Isa 48:18; 50:2; Ezek 32:2; Nah 1:4; Jon 2:4; Hab 3:8; Pss 24:2; 66:6; 72:8; 80:12; 89:26; 98:7-8.

458. E.g., Isa 17:12-13; 23:2-3; 43:16; 51:10; Ezek 27:26; Hab 3:15; Pss 77:20; 93:4; Prov 8:29.

459. The fact that the Punic word here is most likely *ymm* would provide an additional argument against interpreting Poseidon above as Yamm ("Sea") rather than Ba^c^al-Ṣaphon as we have argued. It is quite improbable that "sea(s)" would appear twice in this list, under *Poseidōn* and *limnōn*.

460. Thus I would hold that the alleged parallel cited by Weinfeld ("Loyalty Oath" 397 n. 172), *ʾgmy mym* (Pss 107:35; 114:8) = LXX *limnas hydatōn* ("lakes/pools of water"), though interesting, is not really relevant. This is evident from the fact that neither passage occurs in a cosmogonic or mythological context, whereas *ym(ym) // nhr(wt)* and *ym(ym) // mym (rbym)* are most often found in precisely this sort of context.

461. In Hebrew *rbym* could mean either "many" or "mighty." That the latter is the intended nuance in *mym rbym* is clear from occasional variations of this phrase such as *mym ^c^zym (// ym)*, "*mighty* waters" (Isa 43:16) and *mym zydnym*, "*raging* waters" (Ps 124:5).

462. H. G. May, "Some Cosmic Connotations of *Mayim Rabbîm*, 'Many Waters'," *JBL* 74 (1955) 10.

463. It is possible that "rivers," like "sea" (*ym*), occurs in the Ugaritic "pantheon list(s)." In one line of the text (18) the Akkadian translation reads ḪUR.SAG.MEŠ *u a-mu-tu[m]*. The corresponding Ugaritic is lost here; but in a text listing the same deities in (almost) the same order (609) the corresponding line has *ǵrm* ("mountains"), apparently followed by *w* ("and"; Nougayrol's reading here, [*ǵ*]*rm w*[ ], seems more probable--judging from the copy--than Virolleaud's *ǵrm.š*[ ]--[*Ugaritica V* 580]). Both ḪUR.SAG.MEŠ *u* and *ǵrm w* mean "mountains and. . . ." But what is the second word? Nougayrol (p. 53), concerning Akkadian *a-mu-tu[m]*, notes: "One could actually see in *a-mu-tu*$_4$ an (unprecedented) [*inédit*] plural of *ammu*. . . . [But] the fact remains that *ammu*, 'primordial river (?),' is a very rare word and, up to now, without correspondence in Ugaritic." (The word is attested as a synonym for the Tigris; see *CAD* "A"/2 77a; *AHW* 44b.) It is perhaps relevant that in Hittite treaties "mountains" (ḪUR.SAG.MEŠ) is always followed immediately by "rivers" (ÍD.MEŠ); in several instances (Wl B Rev 29; Wl Rev 58) the two words are connected by *ù*, "and." In 609.6 too the missing word could be *nhrt*, "rivers," since at the end of the word a horizontal wedge, the sign for *t*, is preserved--thus *ǵrm w*[*nhr*]*t*. If the reading "mountains and rivers" is correct, we have another link between the Ugaritic lists and the treaty tradition.

464. May, "*Mayim Rabbîm*" 20.

465. Gen 1:2; Exod 15:8; Ezek 26:19; 31:4; Jonah 2:6; Hab 3:10; Ps 77:17-18.

466. W3 Rev IV.38-40.

467. For the translations that render the phrase in this way, see nn. 428-29 above.

468. "Phoenicia" col. 3749.

469. LSJ 926.

470. Ibid.

471. See Tallqvist, *Akkadische Götterepitheta* 332.

472. Cf. VTE 21, 40; as is clear from fragment 45A (Wiseman, *Vassal-Treaties* pl. 17), "the gods *who dwell in* heaven and earth" (DINGIR.MEŠ *a-ši-bu-ti* AN$^{e}$ [*u* KI$^{tim}$]) is a free variant of "the gods *of* heaven and earth" (DINGIR.MEŠ *šá* AN$^{e}$ *u* KI$^{tim}$).

473. In West Semitic one finds four basic variations of this construction: (1) DN GN (construct chain); (2) DN *b*-GN or DN GN-*h* (with the preposition *b*-, "in," or the locative suffix -*h*); (3) DN *yšb/škn* (*b*-)GN (with the participle, "dwelling in"); (4) DN *b*$^{c}$*l(t)* GN (with *b*$^{c}$*l*/*b*$^{c}$*lt*, "Lord/Lady of").

Examples of (1) include Ugaritic *ʾaṯrt ṣrm* (*CTA* 14[KRT]. 4.201), "Ašerah of Tyre"; Phoenician $^{c}$*štrt kt* (*KAI* 37 A.5), "$^{c}$Aštart of Kition"; and most likely Hebrew *yhwh-m ṣywn* (// *škn yrwšlm*; Ps 135:21), "Yahweh of Zion" (see M. Dahood, *Psalms III: 101-150* [AB 17A; Garden City, N.Y.: Doubleday, 1970] 262-63); cf. $^{d}$*Ištar ša* $^{uru}$*Ninua*.

Examples of (2) include Ugaritic *mlk b*$^{c}$*ṯtrt* (608.17) and *mlk* $^{c}$*ttrth* (607.41), "Milk-in-$^{c}$Aštart"; Phoenician *tnt blbnn* (*KAI* 81.1), "Tanit-in-Lebanon"; Hebrew *yhwh bṣywn* (Ps 99:2: "Great is Yahweh-in-Zion," probably not "Yahweh is great in Zion"; cf. Ps 65:2: *ʾlhym bṣywn*: "O God-in-Zion" [// *šōmēa*$^{c}$ *tĕpillâ*], not "O God, in Zion").

For (3) note Aramaic *yhw ʾlh škn yb* (Kraeling 12.2), "the god Yahu, who dwells in Elephantine"; Hebrew *yhwh yšb ṣywn* (Ps 9:12) and *yhwh škn bṣywn* (Joel 4:21), "Yahweh, who dwells in Zion" (cf. also Pss 2:4; 123:1; 135:21; Deut 33:16; Isa 8:18; 33:5; Joel 4:17); cf. $^{d}$*Ištar āšibat* $^{uru}$*Ninua*.

For (4) note Phoenician *mlqrt b*$^{c}$*l ṣr* (*KAI* 47.1), "Melqart, Lord of Tyre"; *b*$^{c}$*lt.gbl* (*KAI* 5.2), "(DN), Lady of Byblos"; cf. $^{d}$*Ištar bēlet* $^{uru}$*Ninua*.

474. LSJ 1651.

475. Ibid.

476. See BDF §23.

477. See p. 36.

478. *CAD* "K" 210b.

479. J. Friedrich, *Hethitisches Wörterbuch: Kurzgefasste Sammlung der Deutungen hethitischer Wörter* (Heidelberg: Winter, 1952-53) 232a.

480. *DISO* 147.

481. *HALAT* 540b.

482. Cf. an example cited by LSJ (883b): *ta kata polemon, "military* affairs."

483. BAG$^2$ 407-08.

484. Walbank, *Historical Commentary* 50; Paton, *Polybius* 421; Shuckburgh, *Histories* 515.

485. Chroust, "International Treaties" 68; cf. also Egelhaaf, "Analekten" 458 (*welche etwa*); Groag, *Hannibal* 83 (*so viele*).

486. Groag, ibid.; Bickerman, "Hannibal's Covenant" 6; Chroust, "International Treaties" ibid.

487. Paton, *Polybius*; Egelhaaf, "Analekten" (*vorstehen*).

488. Shuckburgh, *Histories* ibid.

489. Weinfeld, "Loyalty Oath" 397.

490. W1 Rev 39,58; W2 Rev 10, 25?

491. For the justification of this restoration (which is actually that of Weidner, *PDK* 48), see the following parallel passages: W1 Rev 58-59; W2 Rev 25; W3 Rev IV.45-46; W4 Rev 10-11.

492. RS 17.79 + 374.54′ (see Kestemont, "Traité" 112).

493. F2 Rev 55; F6 I.40; KBo IV 10 Rev 50; VIII 35 Obv II.9; KUB XXIII 77a Obv 2; XIX 58.17′.

494. Friedrich, *Hethitisches Wörterbuch* 228a.

495. F6 I.39-40.

496. KBo IV 10.50.

497. KUB XIX 58.17′. The same phrase appears with little or no variation in F6 I.39-40; F2 Rev 55; KBo IV 10 Rev 50-51.

498. The phrase *ina libbi* (lit., "in the heart of") is actually equivalent to *ina*, "in"; see *CAD* "L" 174a.

499. W1 Rev 58; see also W2 Rev 10. The *a-na*, "to," in RS 17.357.6′ (where we could expect *i-na*, "in, at," as elsewhere) is probably to be explained as a confusion with the Akkadogram *A-NA*, which marks the dative-locative case in Hittite; cf. [*ke-e-da-ni A-NA*] *NI-IŠ* DINGIR-*LIM* ( = *kēdani linkiya*, "at [the conclusion of] this treaty") in F2 Rev 55.

500. *BBS* no. 25 Rev 25-36.

501. W1 Rev 38-39.

502. For this translation, see A. Goetze in *ANET*$^3$ 205.

503. See LSJ 745b; BAG$^2$ 330b.

504. Hittite *lingai*- and Akkadian *māmītu*, "oath" (also Hittite *išḫiul* and Akkadian *riksu*, "bond") commonly mean "treaty" as well, but strictly speaking Greek *horkos* does not. Nevertheless this word must be translated "treaty" both here and in the first line of H/P. Despite the understanding of *Horkos hon etheto Annibas* as "The oath which Hannibal swore" (Bickerman, "Hannibal's Covenant" 3; Chroust, "International Treaties" 66; Groag, *Hannibal* 83 [*abgelegten*]) or "deposed" (Bickerman, "Hannibal's Covenant" 6), the phrase unmistakably reflects Semitic idiom and hence is to be translated "The *treaty* which Hannibal *made*." Cf. Akkadian *adê šakānu* (lit., "to *place oaths*"); Ugaritic *msmt št* (lit., "to *place* a bond"); Phoenician *krt ʾlt* (lit., "to cut *oaths*"); Hebrew *śym bryt* (lit., "to *place* a bond"); Aramaic *śym ᶜdyʾ* (lit., "to *place oaths*"); Syric *sm tnwy* (lit., "to *place* a contract"). Furthermore, the use of *pros* with the contracting party (*pros Xenophanē* in H/P) reflects the West Semitic preposition *l*, "to," and must be rendered "*with* Xenophanes." Cf. Ugaritic *wtpllm.mlk.r*[*b* . . .] *mṣmt. lnqmd*.[. . .]*št* (*CTA* 64.16-17), "Now Suppiluliūma, the Gr[eat] King, [. . .] has made [. . .] a treaty *with* Niqmadu"; Phoenician *krt.ln.ʾlt ᶜlm* (*KAI* 27 A.8-10), "The Eternal One [i.e., El] has made a covenant *with* us"; Hebrew *ky bryt ᶜwlm śm ly* (2 Sam 23:5), "For he [i.e., Yahweh] has made an everlasting covenant *with* me."

## List of Works Cited


Albright, W. F. *Archaeology and the Religion of Israel*. Baltimore: Johns Hopkins Press, 1968.

________. "Baal-Zephon." *Festschrift Alfred Bertholet zum 80. Geburtstag gewidmet von Kollegen und Freunden*. Ed. W. Baumgartner et al. Tübingen: Mohr, 1950. Pp. 1-14.

________. *From the Stone Age to Christianity: Monotheism and the Historical Process*. Garden City, N.Y.: Doubleday, 1957.

________. "The Kyle Memorial Excavation at Bethel." *BASOR* 56 (1934) 1-15.

________. *The Proto-Sinaitic Inscriptions and Their Decipherment*. HTS 22. Cambridge: Harvard University Press, 1969.

________. *Yahweh and the Gods of Canaan: A Historical Analysis of Two Contrasting Faiths*. Garden City, N.Y.: Doubleday, 1968.

Alcalay, R. *The Complete Hebrew-English Dictionary*. Rev. ed. Bridgeport, Conn.: Prayer Book Press, 1974.

*Ankara Arkeoloji Müzesinde bulunan Boğazköy Tabletleri*. (Boğazköy Tafeln im archäologischen Museum zu Ankara.) Istanbul: Milli eğetim basimevi, 1948.

Arndt, W. F., and Gingrich, F. W., eds. *A Greek-English Lexicon of the New Testament and Other Early Christian Literature*. 2nd. ed. Chicago/London: University of Chicago Press.

Astour, M. C. "Two Ugaritic Serpent Charms." *JNES* 27 (1968) 13-36.

Attridge, H. W., and Oden, R. A., Jr. *Philo of Byblos: The Phoenician History: Introduction, Critical Text, Translation, Notes* (CBQMS 9; Washington: The Catholic Biblical Association of America, 1981).

Avery, C. B., ed. *The New Century Classical Handbook*. New York: Appleton-Century-Crofts, 1962.

Avishur, Y. "Studies of Stylistic Features Common to the Phoenician Inscriptions and the Bible." *UF* 8 (1967 1-22.

Avi-Yonah, M. "Syrian Gods at Ptolemais-Accho." *IEJ* 9 (1959) 1-12.

Barr, J. "Philo of Byblos and His 'Phoenician History'." *BRJL* 57 (1974-1975) 17-68.

Barré, M. L. "$^{d}$LAMMA and Rešep at Ugarit: The Hittite Connection." *JAOS* 98 (1978) 465-67.

Barreca, F. "Lo scavo del tempio." *Ricerche puniche ad Antas: Rapporto preliminare della Missione archeologica dell'Università di Roma e della Soprintendenza alle Antichità di Cagliari.* Ed. E. Acquaro et al. StSem 30. Rome: Istituto di studi del Vicino Oriente; Università di Roma, 1969. Pp. 9-46.

Baudissin, W. W. *Adonis und Esmun: Eine Untersuchung zur Geschichte des Glaubens an Auferstehungsgötter und an Heilgötter.* Leipzig: J. C. Hinrichs, 1911.

Baumgartner, W., Hartmann, B., and Kutscher, E. Y. *Hebräisches und aramäisches Lexikon zum Alten Testament.* 3rd ed. Leiden: Brill, 1967-.

Benedetto, L. F. "Le divinità del giuramento annibalico." *Rivista indo-greco-italica di filologia-lingua-antichità* 3 (1920) 101-25.

Bengtson, H. *Die Staatsverträge des Altertums:* Vol. 2: *Die Verträge der griechisch-römischen Welt von 700 bis 338 v. Chr.* Munich: Beck, 1962.

Benz, F. L. *Personal Names in the Phoenician and Punic Inscriptions: A Catalog, Grammatical Study and Glossary of Elements.* Studia Pohl 8. Rome: Biblical Institute Press, 1972.

Berger, P. "La trinité carthaginoise: Mémoire sur un bandeau trouvé dans les environs de Batna et conservé au musée de Constantine." *Gazette archéologique* 5 (1879) 133-40, 222-29; 6 (1880) 18-31, 164-69.

Bickerman, E. J. "Hannibal's Covenant," *AJP* 73 (1952) 1-23.

________. "An Oath of Hannibal." *TAPA* 75 (1944) 87-102.

Blass, F., and Debrunner, A. *A Greek Grammar of the New Testament and Other Early Christian Literature.* Tr. and ed. R. W. Funk. Chicago: University of Chicago Press, 1961.

Borger, R. *Assyrisch-babylonische Zeichenliste.* AOAT 33. Neukirchen-Vluyn: Neukirchener Verlag, 1978.

________. "Anath-Bethel." *VT* 7 (1957) 102-4.

________. "Der Aufstieg des neubabylonischen Reiches." *JCS* 19 (1965) 57-78.

________. *Die Inschriften Asarhaddons Königs von Assyrien.* AfO Beiheft 9. Osnabrück: Biblio-Verlag, 1967 (reprint of the 1956 edition).

________. "Marduk-zākir-šumi und der Kodex Hammurapi." *Or* 34 (1965) 168-69.

________. Review in *BO* 22 (1965) 166-67 of *CAD* "Z".

________. "Zu den Asarhaddon-Verträgen aus Nimrud." *ZA* 54 [N.F. 20] (1961) 173-96.

Bottéro, J. "Les divinités sémitiques anciennes en Mesopotamie." *Le Antiche Divinità Semitiche*. Ed. S. Moscati. StSem 1. Rome: Centro di studi semitici; Istituto di studi orientali; Università di Roma, 1958. Pp. 17-63.

Bounni, A. "Nabû palmyrénien." *Or* 45 (1976) 46-52.

Brinkman, J. A. *A Political History of Post-Kassite Babylonia: 1158-722 B.C.* AnOr 43. Rome: Pontifical Biblical Institute, 1968.

Cagni, L. *L'Epopea di Erra*. StSem 34. Rome: Istituto di studi del Vicino Oriente; Università di Roma, 1969.

________. *The Poem of Erra*. SANE 1/3. Malibu, Calif.: Undena, 1977.

Caquot, A. "Problèmes d'histoire religieuse." *La Siria nel Tardo Bronzo*. Ed. M. Liverani. Orientis Antiqui Collectio 9. Rome: Centro per le antichità e la storia dell'arte del Vicino Oriente, 1969. Pp. 61-76.

________, Sznycer, M., and Herdner, A. *Textes ougaritiques: Tome I: Mythes et légendes*. LAPO 7. Paris: Editions du Cerf, 1974.

Cary, M., et al., eds. *The Oxford Classical Dictionary*. Oxford: Clarendon Press, 1949.

Caskel, W. "Die alten semitischen Gottheiten in Arabien." *Le Antiche Divinità Semitiche*. Ed. S. Moscati. StSem 1. Rome: Centro di studi semitici; Istituto di studi orientali; Università di Roma, 1958. Pp. 95-117.

Chroust, A.-H. "International Treaties in Antiquity: The Diplomatic Negotiations Between Hannibal and Philip V of Macedonia." *Classica et Mediaevalia* 15 (1954) 60-107.

Clay, A. T., ed. *Babylonian Records in the Library of J. Pierpont Morgan*. 4 vols. Vols. 1-3: New York (privately printed), 1912-1914; vol. 4: New Haven, Conn.: Yale University Press, 1923.

Clemen, C. *Die phönizische Religion nach Philo von Byblos*. MVAG 42/3. Leipzig: J. C. Hinrichs, 1939.

Cogan, M. *Imperialism and Religion: Assyria, Judah and Israel in the Eighth and Seventh Centuries B.C.E.* SBLMS 19. Missoula, Mont.: Scholars Press, 1974.

Coogan, M. D. *West Semitic Personal Names in the Murašû Documents*. HSM 7. Missoula, Mont.: Scholars Press, 1976.

*Corpus Inscriptionum Latinarum: Consilio et Auctoritate Academiae Litterarum Regiae Borussicae editum.* Berlin: Reimerum, 1863-.

*Corpus Inscriptionum Semiticarum: ab Academia Inscriptionum et Litterarum Humaniorum conditum atque digestum.* Paris: e Reipublicae typographeo, 1881-.

Cowley, A. *Aramaic Papyri of the Fifth Century B.C.* Osnabrück: Zeller, 1967 (reprint of the 1923 edition).

Cross, F. M., Jr. *Canaanite Myth and Hebrew Epic: Essays in the History of the Religion of Israel.* Cambridge: Harvard University Press, 1973.

________. "'ēl." *Theological Dictionary of the Old Testament.* Ed. G. J. Botterweck and H. Ringgren. Grand Rapids, Mich." Eerdmans, 1977. 1.242-61.

________. "The Old Phoenician Inscription from Spain Dedicated to Hurrian Astarte." *HTR* 64 (1971) 189-95.

________. "The 'Olden Gods' in Ancient Near Eastern Creation Myths." *Magnalia Dei, the Mighty Acts of God: Essays on the Bible and Archaeology in Memory of G. Ernest Wright.* Ed. F. M. Cross, W. E. Lemke, and P. D. Miller, Jr. Garden City, N.Y.: Doubleday, 1976, Pp. 329-38.

________. "The Origin and Early Evolution of the Alphabet." *Eretz-Israel* 8 (1967) 8*-24*.

________, and Freedman, D. N. *Early Hebrew Orthography: A Study of the Epigraphic Evidence.* AOS 36. New Haven, Conn.: American Oriental Society, 1952.

________, and Saley, R. J. "Phoenician Incantations on a Plaque of the Seventh Century B.C. from Arslan Tash in Upper Syria." *BASOR* 197 (1970) 42-49.

Cumont, F. "Gad." *Pauly-Wissowa*, vol. 7, cols. 433-35.

Dahood, M. *Psalms III: 101-150.* Anchor Bible 17A. Garden City, N.Y.: Doubleday, 1970.

de Moor, J. C. "'ᵃshērāh." *Theological Dictionary of the Old Testament*, vol. 1, 438-44.

de Vaux, R. "Les prophètes de Baal sur le Mont Carmel." *Bulletin du Musée de Beyrouth* 5 (1941) 7-20.

Deller, K., and Parpola, S. "Ein Vertrag Assurbanipals mit dem arabischen Stamm Qedar." *Or* 37 (1968) 464-66.

Dhorme, E., and Dussaud, R. *Les religions de Babylonie et d'Assyrie; les religions des hittites et des hourrites, des phéniciens et des syriens.* Les anciennes religions orientales 2. Paris: Presses universitaires de France, 1949.

Donner, H., and Röllig, W. *Kanaanäische und aramäische Inschriften.* 3 vols. Wiesbaden: Harrassowitz, 1964-1968.

du Mesnil du Buisson, R. *Etudes sur les dieux phéniciens hérités par l'empire romain.* EPROER 14. Leiden: Brill, 1970.

________. *Nouvelles études sur les dieux et les mythes de Canaan.* EPROER 33. Leiden: Brill, 1973.

________. *Les tessères et les monnaies de Palmyre: Un art, une culture et une philosophie grecs dans les moules d'une cité et d'une religion sémitiques.* Paris: Editions E. de Boccard, 1962.

Dumbrell, W. J. "The Tell el-Maskhuta Bowls and the 'Kingdom' of Qedar in the Persian Period." *BASOR* 203 (1971) 33-44.

Dunand, M., and Duru, R. *Oumm el-ᶜAmed: une ville de l'époque hellénistique aux échelles de Tyr.* Paris: A. Maisonneuve, 1962.

Dupont-Sommer, A., and Starcky, J. *Les inscriptions araméens de Sfiré (Stèles I et II).* Paris: Imprimerie nationale, 1958.

Dussaud, R. "Astarté, Pontos et Baᶜal." *CRAIBL* (1947) 201-24.

________. "Melqart." *Syria* 25 (1946-1948) 205-30.

________. *La pénétration des arabes en Syrie avant l'Islam.* Paris: Geuthner, 1955.

Ebeling, E. "Mittelassyrische Rezepte zur Herstellung von wohlriechenden Salben." *Or* 17 (1948) 129-45; 299-313; pls. 1-48.

________. *Stiftungen und Vorschriften für assyrische Tempel.* Berlin: Akademie Verlag, 1954.

Egelhaaf, G. "Analekten zur Geschichte des zweiten punischen Krieges." *Historische Zeitschrift* 53 (1885) 430-69.

Eissfeldt, O. "Baᶜalšamēm und Jahwe." *ZAW* 57 (1939) 1-31.

________. *Baal Zaphon, Zeus Kasios und der Durchzug der Israeliten durchs Meer.* Halle: Niemeyer, 1932.

________. "Der Gott Bethel." *Kleine Schriften.* Ed. R. Sellham and F. Maass. 5 vols. Tübingen: Mohr, 1962. 1.206-33.

Février, J. G. "Paralipomena punica." *Cahiers de Byrsa* 6 (1956) 13-25.

Fisher, L. R., ed. *Ras Shamra Parallels: The Texts from Ugarit and the Hebrew Bible.* Vols. 1-2. AnOr 49-50. Rome: Pontificium Institutum Biblicum, 1972-1975.

Fitzgerald, A. "The Mythological Background for the Presentation of Jerusalem as a Queen and False Worship as Adultery in the Old Testament." *CBQ* 34 (1972) 403-16.

Fitzmyer, J. A. *The Aramaic Inscriptions of Sefîre.* BibOr 19. Rome: Pontifical Biblical Institute, 1967.

________. "The Phoenician Inscription from Pyrgi." *JAOS* 86 (1966) 285-97.

Forrer, E. O. "Eine Geschichte des Götterkönigtums aus dem Hatti-Reiche." *Mélanges Franz Cumont.* AIPHOS 4. Brussels: Sécretariat de l'Institut, 1936. Pp. 687-713.

Frankena, R. *Tākultu: De sacrale maaltijd in het assyrische ritueel: Met een overzicht over de in Assur vereerde goden.* Leiden: Trio, 1953.

________. "The Vassal-Treaties of Esarhaddon and the Dating of Deuteronomy." *OTS* 14 (1965) 122-54.

Friedrich, J. *Hethitisches Elementarbuch.* 2 vols. 2nd ed. Heidelberg: Winter, 1967.

________. *Hethitisches Wörterbuch: Kurzgefasste kritische Sammlung der Deutungen hethitischer Wörter.* Heidelberg: Winter, 1952-1953.

________. *Staatsverträge des Hatti-Reiches in hethitischer Sprache.* Part 1: MVAG 31/1 (1926); Part 2: MVAG 34/1 (1930).

________, and Röllig, W. *Phönizisch-punische Grammatik.* Rev. ed. AnOr 46. Rome: Pontificium Institutum Biblicum, 1970.

Fulco, W. J. *The Canaanite God Rešep.* AOS Essay 8. New Haven, Conn.: American Oriental Society, 1976.

Garbini, G. "Le iscrizioni puniche." *Missione archeologica italiana a Malta: Rapporto preliminare della campagna 1963.* Ed. V. Bonello et al. Rome: Centro di studi semitici, Istituto di studi del Vicino Oriente; Università di Roma, 1964. Pp. 83-96.

________. "Note di epigrafia punica--I." *RSO* 40 (1965) 205-13.

García y Bellido, A. "Deidades semitas en la España antigua." *Sef* 24 (1964) 237-75.

________. *Les religions orientales dans l'Espagne romaine.* EPROER 5. Leiden: Brill, 1967.

Geer, R. M. *Diodorus of Sicily.* Vol. 10. LCL. Cambridge: Harvard University Press, 1954.

Gelb, I. J., et al., eds. *The Assyrian Dictionary of the Oriental Institute of the University of Chicago.* Chicago: The Oriental Institute; Glückstadt: J. J. Augustin, 1964-.

Gibson, J. C. L. *Textbook of Syrian Semitic Inscriptions: Volume II: Aramaic Inscriptions.* Oxford: Clarendon Press, 1975.

Goetze, A. *Die Annalen des Muršiliš.* MVAG 38/6. Leipzig: J. C. Hinrichs, 1933.

________, and Hoffner, H. A., eds. *The Hittite Dictionary of the Oriental Institute of the University of Chicago.* 3/1. Chicago: The Oriental Institute of the University of Chicago, 1980.

Gordon, C. *Ugaritic Textbook.* AnOr 38. Rome: Pontifical Biblical Institute, 1965.

Graziani, S. "Note sui Sibitti." *AION* 39 [n.s. 29] (1979) 673-90.

Greenberg, M. *The Ḫab/piru.* AOS 39. New Haven, Conn.: American Oriental Society, 1955.

Groag, E. *Hannibal als Politiker.* Vienna: L. W. Seidel & Sohn, 1929.

Gröndahl, F. See Quest-Gröndahl, F.

Gsell, S. *Histoire ancienne de l'Afrique du Nord.* 8 vols. Paris: Hachette, 1914-1928.

Gütterbock, H. G. "Hittite Mythology." *Mythologies of the Ancient World.* Ed. S. N. Kramer. Garden City, N.Y.: Doubleday, 1961. Pp. 139-79.

________. "Hittite Religion." *Ancient Religions* (originally published as *Forgotten Religions*). Ed. V. Ferm. New York: Philosophical Library, 1950. Pp. 83-109.

Gurney, O. R. "Hittite Prayers of Mursili II." *AAA* 27 (1940) 3-163.

________. *The Hittites.* London: Penguin, 1954.

________. *Some Aspects of Hittite Religion.* Oxford: Oxford University Press, 1977.

Guzzo Amadasi, M. G. *Le iscrizioni fenicie e puniche delle colonie in occidente.* StSem 28. Rome: Istituto di studi del Vicino Oriente; Università di Roma, 1967.

________. "Note sul dio Sid." *Ricerche puniche ad Antas: Rapporto preliminare della Missione archeologica dell'Università di Roma e della Soprintendenza alle Antichità di Cagliari.* Ed. E. Acquaro et al. StSem 30. Rome: Istituto di studi del Vicino Oriente; Università di Roma, 1969. Pp. 95-104.

Halff, G. "L'onomastique punique de Carthage, répertoire et commentaire." *Karthago* 12 (1965) 61-146.

Harden, D. *The Phoenicians.* Rev. ed. Ancient Peoples and Places 26. London: Thames and Hudson, 1963.

Harper, R. F. *Assyrian and Babylonian Letters Belonging to the K. Collection of the British Museum.* 14 vols. Chicago: University of Chicago Press, 1892-1914.

Harris, Z. S. *A Grammar of the Phoenician Language.* AOS 8. New Haven, Conn.: American Oriental Society, 1936.

Haussig, H. W., ed. *Wörterbuch der Mythologie:* Vol. I: *Götter und Mythen im Vorderen Orient*. Stuttgart: Klett, 1965.

Herdner, A. *Corpus des tablettes en cunéiformes alphabétiques découvertes à Ras Shamra-Ugarit de 1929 à 1939.* MRS 10. Paris: Imprimerie nationale, 1963.

Herrmann, W. "Aštart." *MIO* 15 (1969) 6-52.

Hill, G. F. *A Catalogue of the Greek Coins in the British Museum: vol. 26, Catalogue of the Greek Coins of Phoenicia.* Bologna: Arnaldo Forni, 1965.

________. *A Catalogue of the Greek Coins in the British Museum: vol. 27, Catalogue of the Greek Coins of Palestine (Galilee, Samaria, and Judaea).* Bologna: Arnaldo Forni, 1965.

________. "Some Graeco-Roman Shrines." *JHS* 31 (1911) 56-64.

Hillers, D. R. "Hear, O Heaven, and Give Ear, O Earth." Schaff Lectures, No. 3 (unpublished paper).

________. *Treaty-Curses and the Old Testament Prophets.* BibOr 16. Rome: Pontifical Biblical Institute, 1964.

Hinke, W. J. *A New Boundary Stone of Nebuchadrezzar I. from Nippur.* The Babylonian Expedition of the University of Pennsylvania; Series D: Researches and Treatises 4. Philadelphia: University of Pennsylvania Press, 1907.

Hoffner, H. A., Jr. "The 'City of Gold' and the 'City of Silver'." *IEJ* 19 (1969) 178-80.

Hvidberg-Hansen, F. O. *La Déesse TNT: Une étude sur la religion canaanéo-punique.* Habilitationsschrift der Universität Kopenhagen. Kopenhagen, 1979.

Hyatt, J. P. "The Deity Bethel and the Old Testament." *JAOS* 59 (1939) 81-98.

Jastrow, M. *A Dictionary of the Targumim, the Talmud Babli and Yerushalmi, and the Midrashic Literature.* New York: Jastrow, 1967.

________. *Die Religion Babyloniens und Assyriens.* 2 vols. Giessen: Töpelmann, 1905.

Jean, C.-F., and Hoftijzer, J. *Dictionnaire des inscriptions sémitiques de l'Ouest.* Leiden: Brill, 1965.

Jirku, A. "Zweier-Gottheit und Dreier-Gottheit im altorientalischen Palästina-Syrien." *MUSJ* 45 (1969) 399-404.

Joüon, P. P. *Grammaire de l'hébreu biblique.* Rome: Institut Biblique Pontifical, 1923.

Katzenstein, H. J. *The History of Tyre: From the Beginning of the Second Millenium* [sic] *B.C.E. until the Fall of the Neo-Babylonian Empire in 538 B.C.E.* Jerusalem: The Schocken Institute for Jewish Research, 1973.

Kaufman, S. A. *The Akkadian Influence on Aramaic.* Assyriological Studies 19. Chicago/London: University of Chicago Press, 1974.

*Keilschrifttexte aus Boghazköi.* Leipzig: J. C. Hinrichs, 1916-1974; Berlin: Gebr. Mann, 1976.

*Keilschrifturkunden aus Boghazköi.* Berlin: Akademie Verlag, 1911-.

Kempinski, A., and Košak, S. "Der Išmeriga Vertrag." *WO* 5 (1970) 191-217.

Kestemont, G. "Le panthéon des instruments hittites de droit public." *Or* 45 (1976) 147-77.

________. "Le traité entre Mursil II de Hatti et Niqmepa d'Ugarit." *UF* 6 (1974) 85-127.

King, L. W. *Babylonian Boundary-Stones and Memorial-Tablets in the British Museum.* London: Oxford University Press, 1912.

Klengel, H. "Der Wettergott von Ḫalab." *JCS* 19 (1965) 87-93.

Komoroczy, G. "Das Pantheon im Kult, in den Götterlisten und in der Mythologie." *Or* 45 (1976) 80-86.

Korošec, V. "Die Götteranrufung in den keilschriftlichen Staatsverträgen." *Or* 45 (1976) 120-29.

________. *Hethitische Staatsverträge: Ein Beitrag zu ihrer juristischen Wertung.*

Kraeling, E. *The Brooklyn Museum Aramaic Papyri: New Documents of the Fifth Century B.C. from the Jewish Colony at Elephantine.* New Haven, Conn.: Yale University, 1953.

Kümmel, H. M. "Die Religion der Hethiter." *Theologie und Religionswissenschaft.* Ed. U. Mann. Darmstadt: Wissenschaftliche Buchgesellschaft, 1973. Pp. 65-85.

Labat, R., and Malbran-Labat, F. *Manuel d'épigraphie akkadienne (Signes, Syllabaire, Idéogrammes).* Rev. ed. Paris: Geuthner, 1976.

________. "Elam c. 1600-1200 B.C." *The Cambridge Ancient History.* 2/2. *History of the Middle East and the Aegean Region c. 1380-1000 B.C.* Ed. I. E. S. Edwards et al. 3rd edition. Cambridge: Cambridge University Press, 1975. Pp. 379-416.

Lambert, W. G. "The Historical Development of the Mesopotamian Pantheon: A Study in Sophisticated Polytheism." *Unity and Diversity: Essays in the History, Literature, and Religion of the Ancient Near East.* Ed. H. Goedicke and J.J.M. Roberts. Baltimore: Johns Hopkins University Press, 1975. Pp. 191-200.

________. "The Reign of Nebuchadnezzar I: A Turning Point in the History of Ancient Mesopotamian Religion." *The Seed of Wisdom: Essays in Honour of T. J. Meek*. Ed. W. S. McCullough. Toronto: University of Toronto Press, 1964. Pp. 3-13.

Landsberger, B. *Sam'al: Studien zur Entdeckung der Ruinenstätte Karatepe*. Ankara: Türkische historische Gesellschaft, 1948.

Langdon, S. "A Phoenician Treaty of Assarhaddon: Collation of K.3500." *RA* 26 (1929) 189-94.

________. *Tammuz and Ishtar: A Monograph upon Babylonian Religion and Theology*. Oxford: Clarendon Press, 1914.

Laroche, E. "Les dénominations des dieux 'antiques' dans les textes hittites." *Anatolian Studies Presented to Hans Gustav Güterbock on the Occasion of his 65th Birthday*. Ed. K. Bittel et al. İstanbul: Nederlands historisch-archaeologisch Instituut in het Nabije Oosten, 1974. Pp. 175-85.

________. "Panthéon national et panthéons locaux chez les Hourrites." *Or* 45 (1976) 94-99.

________. "Recherches sur les noms des dieux hittites." *RHA* 7 (1946-1947) 7-139 (also published separately).

Levi della Vida, G. Review in *RSO* 39 (1964) 314-20 of *Missione archeologica italiana a Malta: Rapporto preliminare della campagna 1963*. Ed. V. Bonello et al. Rome: Centro di studi semitici, Istituto di studi del Vicino Oriente; Università di Roma, 1964.

Lewis, C. T., and Short, C. *A Latin Dictionary: Founded on Andrews' Edition of Freund's Latin Dictionary* Oxford: Clarendon Press, 1879.

L'Heureux, C. E. *Rank Among the Canaanite Gods El, Ba$^{c}$al, and the Repha'im*. HSM 21. Missoula, Mont.: Scholars Press, 1979.

Liddell, H. G., Scott, R., and Jones, H. S. *A Greek-English Lexicon (with Supplement)*. Rev. ed. Oxford: Clarendon Press, 1968.

Lidzbarski, M. *Ephemeris für semitische Epigraphik*. 3 vols. Giessen: Rickers-Töpelmann, 1902-1915.

Lipiński, E. "La fête de l'ensevelissement et de la résurrection de Melqart." *Actes de la XVII$^{e}$ Rencontre assyriologique internationale: Bruxelles, 30 juin-4 juillet 1969*. Ed. A. Finet. Ham-sur-Heure: Comité belge de recherches en Mesopotamie, 1970. Pp. 30-58.

Liverani, M. "La preistoria dell'epiteto 'Yahweh ṣĕbā'ōt'." *AION* 17 (1967) 331-34.

________. *Storia di Ugarit nell'età degli archivi politici*. StSem 6. Rome: Istituto di studi del Vicino Oriente; Università di Roma, 1962.

________. Review in *OrAnt* 8 (1969) 337-40 of *Ugaritica V*.

Luckenbill, D. D. *The Annals of Sennacherib.* OIP 2. Chicago: University of Chicago Press, 1924.

May, H. G. "Some Cosmic Connotations of *Mayim Rabbîm*, 'Many Waters'." *JBL* 74 (1955) 9-21.

McCarthy, D. J. *Treaty and Covenant: A Study in Form in the Ancient Oriental Documents and in the Old Testament.* AnBib 21. Rome: Pontifical Biblical Institute, 1963.

________. *Treaty and Covenant: A Study in Form in the Ancient Oriental Documents and in the Old Testament.* 2nd edition. AnBib 21A. Rome: Pontifical Biblical Institute, 1978.

Meltzer, O., and U. Kahrstedt. *Geschichte der Karthager.* 3 vols. Berlin: Weidmann, 1879-1913.

Mendenhall, G. E. "The Ancient in the Modern--and Vice Versa." *Michigan Oriental Studies in Honor of George G. Cameron.* Ed. L. L. Orlin. Ann Arbor, Mich.: University of Michigan Press, 1976. Pp. 227-53.

Merlin, A. *Le sanctuaire de Baal et de Tanit près de Siagu.* Notes et documents publiés par la direction des antiquités et arts 4. Paris, 1910.

Messerschmidt, L., and Ungnad, A. *Vorderasiatische Schriftdenkmäler der Königlichen Museen zu Berlin* 1. Leipzig, 1907.

Meyer, E. "Baal." *Ausführliches Lexikon der griechischen und römischen Mythologie.* Ed. W. H. Roscher. Leipzig: Teubner, 1884-1937. 1/2. cols. 2867-80.

________. "Phoenicia." *Encyclopedia Biblica: A Critical Dictionary of the Literary, Political, and Religious History, Archaeology, Geography, and Natural History of the Bible.* 4 vols. Ed. T. K. Cheyne and J. S. Black. New York: Macmillan, 1902. 3.cols. 3730-65.

Milik, J. T. "Les papyrus araméens d'Hermoupolis et les cultes syro-phéniciens en Egypte perse." *Bib* 48 (1967) 546-622.

Miller, P. D., Jr. *The Divine Warrior in Early Israel.* HSM 5. Cambridge: Harvard University Press, 1973.

Moore, J. M. *The Manuscript Tradition of Polybius.* Cambridge: Cambridge University Press, 1965.

Moscati, S. "Astarte in Italia." *Rivista di cultura classica e medioevale* 7 (1965) 756-60.

________. *I Fenici e Cartagine.* Turin: Unione tipografico-editrice torinese, 1972.

________., ed. *Le Antiche Divinità Semitiche.* StSem 1. Rome: Centro di studi semitici; Istituto di studi orientali; Università di Roma, 1958.

________. *The World of the Phoenicians.* Tr. by A. Hamilton. London: Weidenfeld and Nicolson, 1968.

Movers, F. C. *Die Phönizier*. 3 vols. Bonn: Weber, 1841.

Mras, K., ed. *Eusebius Werke:* Vol. 1, Part 8: *Die Praeparatio evangelica*. Berlin: Akademie-Verlag, 1954.

Münter, F. *Religion der Karthager*. Copenhagen: Schubothe, 1821.

Na'aman, N. "Looking for KTK." *MIO* 9 (1977-1978) 220-39.

North, R. "Separated Spiritual Substances in the Old Testament." *CBQ* 29 (1967) 419-49.

Nougayrol, J. *Le Palais royal d'Ugarit III: Textes accadiens et hourrites des Archives Est, Ouest et Centrales*. MRS 6 Paris: Imprimerie nationale, 1955.

________. *Le Palais royal d'Ugarit IV: Textes accadiens des Archives Sud (Archives internationales)*. MRS 9. Paris: Imprimerie nationale, 1956.

________, et al., eds. *Ugaritica V: Nouveaux Textes accadiens, hourrites et ugaritiques des archives et bibliothèques privées d'Ugarit: Commentaires des textes historiques (Première partie)*. MRS 16. Paris: Imprimerie nationale, 1968.

________. "Sirrimu (non **purîmu*) 'âne sauvage'." *JCS* 2 (1948) 203-08.

Oden, R. A., Jr. "Ba$^{c}$al Šamēm and 'Ēl." *CBQ* 39 (1977) 457-73.

________. "The Persistence of Canaanite Religion." *BA* 39 (1976) 31-36.

________. *Studies in Lucian's "De Syria Dea"*. HMS 15. Missoula, Mont.: Scholars Press, 1977.

Oettinger, N. *Die Militärischen Eide der Hethiter*. StBoT 22. Wiesbaden: Harrassowitz, 1976.

Oppenheim, A. L. *Letters From Mesopotamia*. Chicago/London: University of Chicago Press, 1967.

Otten, H. "Eine Beschwörung der Unterirdischen aus Boğazköy." *ZA* 54 [N.F. 20] (1961) 114-57.

________. "Die Götter Nupatik, Pirinkir, Ḫešue und Ḫatni-Pišaišaphi in den hethitischen Felsreliefs von Yazilikaya." *Anatolia* 4 (1959) 27-37.

________. "Die Gottheit Lelvani der Boğazköy-Texte." *JCS* 4 (1950) 119-36.

________. "Ein kanaanäischer Mythus aus Boğazköy." *MIO* 1 (1953) 125-50.

________, and von Soden, W. *Das akkadisch-hethitische Vokabular: KBo I 44 + KBo XIII 1*. StBoT 7. Wiesbaden: Harrassowitz, 1968.

Parpola, S. "The Alleged Middle/Neo-Assyrian Irregular Verb **naṣṣ* and the Assyrian Sound Change š > s." *Assur* 1/1 (1974) 1-10.

Paton, W. R. *Polybius: The Histories.* 6 vols. LCL. Cambridge: Harvard University Press, 1923.

Paul, S. M. "Jerusalem--A City of Gold." *IEJ* 17 (1967) 259-63.

________. "Jerusalem of Gold--A Song and an Ancient Crown." *BAR* 3 (1977) 38-41.

Payne Smith, J. *A Compendious Syriac Dictionary: Founded upon the Thesaurus Syriacus of R. Payne Smith, D.D.* Oxford: Clarendon Press, 1903.

Peck, H. T., ed. *Harper's Dictionary of Classical Literature and Antiquities.* New York: Cooper Square, 1962.

Pettinato, G. "The Royal Archives of Tell Mardikh-Ebla." *BA* 39 (1976) 44-52.

Picard, C. G., and Picard, C. *The Life and Death of Carthage: A Survey of Punic History and Culture from its Birth to its Final Tragedy.* Tr. D. Collon. New York: Taplinger, 1968.

________. *La vie quotidienne à Carthage au temps d'Hannibal.* Paris: Hachette, 1958.

Picard, G. Ch.-. "Carthage au temps d'Hannibal." *Annuario dell'Accademia etrusca di Cortona* 12 (1961-1964) 9-36.

________. *Hannibal.* Paris: Hachette, 1967.

________. *Les religions de l'Afrique antique.* Paris: Plon, 1954.

Pietschmann, R. *Geschichte der Phönizier.* Berlin: Grote, 1889.

Pope, M. *El in the Ugaritic Texts.* VTSup 2. Leiden: Brill, 1955.

Porten, B. *Archives from Elephantine: The Life of an Ancient Jewish Military Colony.* Berkeley/Los Angeles: University of California Press, 1968.

Postgate, J. N. *Fifty Neo-Assyrian Legal Documents.* Warminister: Aris and Phillips, 1976.

________. *Neo-Assyrian Royal Grants and Decrees.* Studia Pohl, series major 1. Rome: Pontifical Biblical Institute, 1969.

Pritchard, J. B. *The Ancient Near East in Pictures Relating to the Old Testament.* 2nd ed. with supplement. Princeton: Princeton University Press, 1969.

________. *Ancient Near Eastern Texts Relating to the Old Testament.* 3rd ed. with supplement. Princeton: Princeton University Press, 1969.

________. *Recovering Sarepta, A Phoenician City: Excavations at Sarafand, Lebanon, 1969-1974, by the University Museum of the University of Pennsylvania.* Princeton: Princeton University Press, 1978.

Quest-Gröndahl, F. *Die Personennamen der Texte aus Ugarit.* Studia Pohl 1. Rome: Päpstliches Bibelinstitut, 1967.

Reiske, I. I. *Animadversionum ad Graecos auctores volumen quartum quo Polybii reliquiae pertractantur.* Leipzig, 1763.

*Répertoire d'épigraphie sémitique.* Paris: Imprimerie nationale, 1900-.

Riess, E. "Ammon." *Pauly-Wissowa*, vol. 1, cols. 1853-58.

Roberts, J.J.M. "The Davidic Origin of the Zion Tradition." *JBL* 92 (1973) 329-44.

________. *The Earliest Semitic Pantheon: A Study of the Semitic Deities Attested in Mesopotamia before Ur III.* Baltimore: Johns Hopkins University Press, 1972.

________. "El." *The Interpreter's Dictionary of the Bible: Supplementary Volume.* Nashville: Abingdon Press, 1976. Pp. 255-58.

Roeder, G. *Urkunden zur Religion des alten Ägyptens.* Jena: Diederichs, 1915.

Saggs, H. W. F. *The Greatness That Was Babylon: A Sketch of the Ancient Civilization of the Tigris-Euphrates Valley.* New York/Toronto: New American Library, 1962.

Scheil, F. V. "Sin-šar-iškun fils d'Aššurbanipal." *ZA* 11 (1896) 47-49.

Schmitt, H. H. *Die Staatsverträge des Altertums:* Vol. 3: *Die Verträge der griechisch-römischen Welt von 338 bis 200 v.Chr.* Munich: Beck, 1969.

Schweighäuser, J. G. *Polybii Megalopolitani historiarum quidquid superest.* 9 vols. Leipzig: Weidmann, 1972.

Seyrig, H. "Le culte de Bêl et de Baalshamîn à Palmyre." *Antiquités syriennes.* Paris, 1934.

Shuckburgh, E. S. *The Histories of Polybius.* New York: Macmillan, 1889.

Solá Solé, J. M. "Inscripciones fenicias de la península ibérica." *Sef* 15 (1955) 41-53.

________. "Miscelanea púnico-hispana I." *Sef* 16 (1956) 325-55.

________. "La plaquette en bronze d'Ibiza." *Sem* 4 (1951-1952) 25-31.

Steinmetzer, F. X. *Die babylonischen Kudurru (Grenzsteine) als Urkundenform.* Studien zur Geschichte und Kultur des Altertums 11/4-5. Paderborn: F. Schöningh, 1922.

Streck, M. *Assurbanipal und die letzten assyrischen Könige bis zum Untergange Niniveh's.* 3 vols. Vorderasiatische Bibliothek 7. Leipzig: J. C. Hinrichs, 1916.

Tallqvist, K. *Akkadische Götterepitheta.* StOrFen 7. Leipzig: Harrassowitz, 1938.

Teixidor, J. "A Note on the Phoenician Inscription from Spain." *HTR* 68 (1975) 197-98.

________. *The Pagan God: Popular Religion in the Greco-Roman Near East.* Princeton: Princeton University Press, 1977.

________. *The Pantheon of Palmyra.* EPROER 79. Leiden: Brill, 1979.

Vassel, E. "Le panthéon d'Hannibal." *Revue tunisienne* 19 (1912) 329-45; 20 (1913) 29-45; 212-19, 307-14, 447-63, 576-79, 654-57; 21 (1914) 48-55, 164-84.

Vattioni, F. "Il dio Resheph." *AION* 15 (1965) 39-74.

Veyne, P. "'Tenir un buste': une intaille avec le génie de Carthage, et le sardonyx de Livie à Vienne." *Cahiers de Byrsa* 8 (1958-1959) 61-78.

von Schuler, E. *Die Kaškäer.* Berlin: de Gruyter, 1965.

von Soden, W. *Akkadisches Handwörterbuch: Unter Benutzung des lexikalischen Nachlasses von Bruno Meissner (1868-1947).* Wiesbaden: Harrassowitz, 1965-.

________. "Die Schutzgenien Lamassu und Schedu in der babylonisch-assyrischen Literatur." *Baghdader Mitteilungen* 3 (1967) 148-56.

Walbank, F. W. *A Historical Commentary on Polybius.* 2 vols. Oxford: Clarendon Press, 1957-1967.

________. *Polybius.* Berkeley/Los Angeles: University of California Press, 1972.

Waser, O. "Tyche." *Ausführliches Lexikon der griechischen und römischen Mythologie.* Ed. W. H. Roscher. Leipzig: Teubner, 1884-1937. 5. cols. 1309-80.

Waterman, L. *Royal Correspondence of the Assyrian Empire: Translated into English, with a Transliteration of the Text and a Commentary* 4 vols. University of Michigan Humanistic Series 17-20. Ann Arbor, Mich.: University of Michigan, 1930.

Wehr, H. *A Dictionary of Modern Written Arabic.* Ed. J. M. Cowan. 3rd ed. Ithaca, N.Y.: Spoken Language Services, 1971.

Weidner, E. F. *Politische Dokumente aus Kleinasien: Die Staatsverträge in akkadischer Sprache aus dem Archiv von Boghazköi.* Boghazköi-Studien 8-9. Leipzig: J. C. Hinrichs, 1923.

________. "Der Staatsvertrag Aššurniraris VI. [*sic*] von Assyrien mit Mati'ilu von Bît-Agusi." *AfO* 8 (1932-1933) 17-34.

________. "Die Feldzüge Šamši-Adads V. gegen Babylonien." *AfO* 9 (1933-1934) 89-104.

Weinfeld, M. "*hšbᶜt hwsʾlym šl ʾsrḥdwn mlk ʾšwr*" ("The Vassal-Treaties of Esarhaddon--an Annotated Translation") *Shnatôn* 1 (1975) 89-122.

________. "The Loyalty Oath in the Ancient Near East." *UF* 8 (1975) 379-414.

________. "*l^cnyyn mwnhy bryt bywnyt wbrwmyt*" ("Greek and Roman Covenantal Terms and Their Affinities to the East") *Leš* 38 (1974) 231-37.

________. "*šbw^ct 'mwnym l'srḥdwn--'wpyyh wmqbylwtyh b^cwlm hmzrḥ hqdwm*" ("The Loyalty Oath in the Ancient Near East") *Shnatôn* 1 (1975) 51-88.

Weippert, M. *Die Landnahme der israelitischen Stämme in der neueren wissenschaftlichen Diskussion.* FRLANT 92. Göttingen: Vandenhoeck & Ruprecht, 1967.

Winckler, H. *Altorientalische Forschungen: 1. Reihe.* Leipzig: Pfeiffer, 1893-1897.

Wiseman, D. J. *The Vassal-Treaties of Esarhaddon* London: British School of Archaeology in Iraq, 1958 ( = *Iraq* [1958] Part I).

________, ed. *The Alalakh Tablets.* Occasional Publications of the British Institute of Archaeology at Ankara 2. London: The British Institute of Archaeology at Ankara, 1953.

________. *Chronicles of Chaldean Kings (626-556 B.C.) in the British Museum.* London: British Museum Publications, 1956.

Wohl, H. "Niraḫ or Šaḫan?" *JANESCU* 5 (1973) 443-44.

Xella, P. "A proposito del giuramento annibalico." *OrAnt* 10 (1971) 189-93.

Yadin, Y. "Symbols of Deities at Zinjirli, Carthage and Hazor." *Near Eastern Archaeology in the Twentieth Century.* Ed. J. A. Sanders (Nelson Glueck Festschrift). Garden City, N.Y.: Doubleday, 1970. Pp. 199-231.

Zadok, R. "Phoenicians, Philistines, and Moabites in Mesopotamia." *BASOR* 230 (1978) 57-63.

Zevit, Z. "A Phoenician Inscription and Biblical Covenant Theology." *IEJ* 27 (1977) 110-18.

Ziebarth, E. *De iureiurando in iure Graeco quaestiones.* Göttingen: Kaestner, 1892.

Ziegler, K. "Tyche." *Pauly-Wissowa,* vol. 14, cols. 1643-96.

Zorell, F. *Lexicon Hebraicum et Aramaicum Veteris Testamenti.* Rome: Pontificium Institutum Biblicum, 1968.

INDEXES

## Deities and Deified Elements

Gods of the army/camp, the, 10, 32-33, 109, 124, 155 n.153
Gods of the Lulaḫḫi/Hapiri, the, 33, 36, 87, 122, 154 n.118, 157 n.174
Gods of the mercenaries, the, 34, 36, 88, 102, 109, 122
Great Anu, the, 44, 161 n. 42
Great gods, the, 19, 21, 24, 105-6, 110-11, 113-14, 116-18, 128-35, 138 n.39
Gula, 9, 24, 54, 107, 111, 113, 118, 133-34, 149 n.67, 150 n.71

Hadad/Haddu, 38, 40, 49-50, 52-54, 56, 62, 81-83, 85, 124, 159 n.11, 169 n.154, 173 n.221, 179 n.345, 181 n.390, 182 n.400. *See also* (H)adad of Aleppo; Baʿal
(H)adad of Aleppo, 25-26, 107-8, 149 n.69, 150 n.69
Ḫaldia, 45
Ḫallara of Dunna, 121
Ḫantitaššu of Hurma, 33, 121
Ḫapantaliya, 119
Ḫazzai, 31, 119
Heaven, 26-28, 35, 57, 89-91, 108, 122, 180 n.374, 184 n.439
Ḫepat, 31, 33, 120, 168 n.154
Hephaestus, 84-85
Hera, 5, 10-11, 13, 18, 37, 40, 42, 57-59, 66, 73, 79, 109, 123, 125, 169 nn.159-60, 178 n.325. *See also* Tanit
Herakles, 10, 62, 64, 74, 76-78, 123, 125, 179 n.345. *See also* Melqart
*Hercules Gaditanus*, 78
Hermes, 13
Ḫešue, 33
High gods, 8-9, 22, 28, 32, 35-37, 73, 80, 83, 86-88, 100-102, 109, 118, 153 n.112, 168 n.157, 180 n.374
Ḥoron, 26
Horus, 70
Ḫumḫummu, 107
Ḫurri, 119, 153 n.110
Ḫuwaššanna of Ḫupišna, 121

Iariqi, 126
Il-aba, 105, 169 n.154
*'Il'Ib*, 49, 165 n.93
$^{d}$IM, 133-34. *See also* Storm-god, the
Innara, 154 n.127
In-Šušinak, 44
Iolaos, 10, 64, 74, 77-79, 123, 125, 179 n.345
Išḫara, 31, 33, 35, 120, 157 n.170
$^{d}$IŠKUR, 157 n.168
Ištar, 8, 31-33, 35, 54, 66, 69, 105-6, 120, 130-31, 136 n.22, 153 n.114, 157 n.170, 165 n.97, 169 n.154
Ištar, the Assyrian, 110, 146 n.36
Ištar of Arbela, 22, 75, 107, 110-11, 113-15, 118, 133, 146 n.35
Ištar of Nineveh, 22, 75, 94, 110-11, 113, 115, 118, 120, 133, 186 n.473
Ištaran. *See* Great Anu, the
Išum, 107
Iyarri, 121

Juno, 10, 58-59, 66, 169 n.159. *See also* Hera; Tanit
Jupiter (planet), 5-6, 17, 22, 112, 116, 126 n.11

Karzi, 119
Kataḫḫa of Ankuwa, 121
*Kd'h*, 24-25, 108
Kippat-mati, 110
Kore, 12
Koṯar(-and-Ḫasis), 26, 84. *See also* Kušor
Kronos, 12-13, 43, 50, 56-57, 63, 74, 79, 159 n.15, 160 n.33. *See also* Baʿal-Ḥam(m)on; Saturn
Kulitta, 33, 120
Kumarbi/Kumurwi, 29, 152 n.95; 159 n.11
Kuniyawanni of Landa, 121
Kušor, 55, 84-86, 124-25, 182 n.417. *See also* Koṯar (-and-Ḫasis); Baʿal-Malage; Triton.
Kušuḫ, 159 n.11

Lady of Byblos, the, 41, 186 n.473
"Lady" of Landa, 121
$^{d}$LAMMA, 31-32, 35, 37, 119, 157 n.169, 158 n.181. *See also* Protective deity
$^{d}$LAMMA of Hatti, 66, 119, 154, 127, 158 n.181
Laṣ, 197-8
Lelwani, 32-33, 120. *See also* Ereškigal/$^{d}$EREŠ.KI. GAL; Sun-goddess of the netherworld, the
Lulutašši, 153 n.109

Place Names

93-94, 123-24

Ḫabban, 138 n.39. *See also* Bît-Ḫabban
Hamath, 40
Ḥam(m)on, 75, 178 n.327
Ḫanaḫana, 120
Ḫapiri, the, 33-34, 155 n.150. *See also* Lulaḫḫi, the
Harran, 113
Hatra, 51, 166 n.106
Ḫattarina, 120
Ḫatti, 7, 9, 16-17, 27, 35, 66, 73, 88, 94, 102, 121-22, 149 nn.67, 69, 154 nn.118, 127, 158 n.181, 159 n.11, 168 n.154
Ḫayasa, 17
Ḫupišna, 121
Ḫurma, 33, 121

Ibiza, 67
Išmeriga, 17, 157 nn.168, 172
Izai, 131

Jebel el-Aqra$^{c}$. *See* (Mount) Ṣaphon
Jerusalem, 186 n.473
Judah, 49

Kakzi, 113
Karatepe, 40-41, 52-53, 56
Karzitali, 131
Kaška/Kaškeans, the, 17, 35, 157 n.168
Katapa, 121
Kinza (= Kadesh), 121
Kition, 165 n.97, 186 n.473
Kizzuwatna/i, 94, 120, 158 n.11
KTK, 23-24, 29, 93, 108-9, 140 n.23, 148 n.58, 149 n.67, 168 n.154

Lagaš, 7
Lake Gennesaret. *See* Sea of Galilee
Landa, 121
Lebanon, 68, 177 n.325, 186 n.473
Leptis Magna, 78
Libya(ns), 76-77
Lulaḫḫi, the, 34, 154 n.118, 155 n.150, 156 n.150. *See also* Ḫapiri, the

Macedonia, 1, 10-11, 14, 16, 18, 36, 87, 94, 97, 109, 123-24
Malac(h)a. *See* Málaga
Málaga, 85-86
Malta, 59, 160 n.35
Media, 131
Memphis, 84
Mesopotamia, 7, 9, 16, 23, 47-49, 55, 66
Mount Casius/Kasios. *See* (Mount) Ṣaphon
Mount Lebanon, 121
Mount Pišaiša, 122
(Mount) Ṣaphon, 81-82
Mount Šariyana, 122
Muṣaṣir, 45

Nahr-el-Kalb, 132
Naḫšimarti, 131
Namar, 133-34
Nineveh, 22, 75, 94, 107, 110-11, 113, 115, 118, 120, 133, 186 n.473
Nippur, 113
Nuḫaššši, 94
Orontes, 71

Palestine, 43, 54, 69
Palmyra (Tadmor), 13, 52, 63, 65-66, 86
Peraia, 63
Phoenicia, 43, 57, 63, 80, 83, 85-86, 178 n.325

Qarqar, 149 n.69
Qedar, 16; 130

Ras Shamra, 84. *See also* Ugarit
RḤBH, 24-25, 108, 148 n.64, 149 n.67
Rome, 1, 13, 100, 143 n.25

Šam'al (Zinčirli) 9, 40, 50, 53, 56, 62, 158 n.11, 159 n.11
Šamuḫa, 33, 121
Ṣaphon. *See* (Mount) Ṣaphon
Sardinia, 76, 179 nn.344-45
Sarepta, 59, 171 n.186
Sea of Galilee, 91
Sefîre, 15-16, 47
Sicily, 74
Sidi-Daoud, 71
Sidon, 84, 171 n.197
Sikrisu, 131
Soleis, 81
Spain, 68, 85
Sumer, 113
Susa, 44
Syria, 9, 40-41, 63, 149 n.67, 158 n.11, 166 n.103, 177 n.325, 178 n.325
Syria-Palestine, 13, 16, 39, 41, 51, 53, 56, 69

Taḫurpa, 121
Tarsus, 173 n.240

Tas Silg, 59
Tawiniya, 120
Thebes, 74
Thinissut, 68, 70
"Trans-Euphrates," 43, 111, 135
Tunisia, 71
Turmitta, 120
Tyre, 16, 20, 38, 41-45, 49-51, 56, 58-59, 62-63, 74, 100, 102, 135-36, 160 n.35, 163 n.72, 165 n.97, 170 n.177, 186 n.473

Uda, 120
Ugarit, 29, 34, 39-40, 48-51, 55, 58-63, 69, 72, 75, 80-82, 84-85, 89, 91, 149 nn.67, 69, 150 nn.69, 75, 152 n.95, 159 n.11, 163 n.73, 164 n.92, 165 n.93, 166 n.103, 169 nn.158, 165, 170 n.176, 178 n.326, 181 n.390

Umm el-ᶜAwāmīd, 43, 78, 160 n.35
Ur, 169 n.154
Urakazabanu, 5-6, 131
Urartu, 45

Ya'udi, 9. *See also* Šam'al
Yazilikaya, 72, 168 n.154

Zamua, 131
Zinčirli, 132. *See also* Šam'al
Zion, 186 n.473

Personal Names

Adad-nirari III, 137 n.23
Aḫatmilku, 72-73
Alexander (the Great), 62
Antiochus IV Epiphanes, 41
Arnuwanda I, 17
Assurbanipal, 16, 116, 130, 161 n.47
Assur-nirari V, 16, 141 n.39
Baᶜal of Tyre, 16; 43
Barcides, the, 12-14, 62, 101, 143 n.25
Barmocar, 6, 14, 178 n.337
Bir-Ga'yah, 29, 108, 140 n.23, 144 n.6, 148 n.58
Bir-Hadad, 163 n.72
Bir-Rakib, 9

Cleomachus, 5

Dido, 12, 64

Esarhaddon, 5-6, 16, 43-44, 131-33, 138 n.30, 141 n.39, 152 n.101
Esther, 73
Eutychides of Sicyon, 71-72

Ḥamilkat, 41
Hammurapi, 145 n.27
Hannibal, 1, 3, 5-6, 10-14, 40, 60, 66, 68, 87, 100-101, 143 n.25, 188 n.504
Hanno of Carthage, 81, 181 n.389
Hattusilis III, 149 n.69, 158 n.11
Hazail, 44, 161 n.47
Hiram of Tyre, 43
Ḫuqqana of Ḫayasa, 17

Kilamuwa, 158 n.11

Mago, 6, 13
Marduk-zakir-šumi I, 16
Mati'ilu/Matiᶜel, 16, 19, 29, 108, 130, 140 n.23, 141 n.39, 145 n.33, 150 n.22
Mursilis II, 9, 141 n.39, 168 n.154
Myrcan, 6, 14

Nabu-apli-iddina, 17, 116, 129, 140 n.23
Nebuchadnezzar (Nabu-kudurruuṣur) I, 18, 133
Niqmadu, 188 n.504

Panamuwa, 9
Philip V of Macedonia, 1, 5, 10-11, 14

Ramatay(a), 5-6
Ramses II, 149 n.69
Ritti-Marduk, 133

Šalmaneser III, 149 n.69
Šamaš-šum-ukin, 129
Šamši-Adad V, 16
Sargon II, 44-45, 132
Scipio, Q. Caecilius Metellus Pius, 68, 70
Seleucids, the, 41
Seleucus Nicator, 71
Sennacherib, 16, 44, 125 n.2, 132, 146 n.35, 149 n.69, 161 n.47
Šilḫak-In-Šušinak, 44
Sin-šum-lišir, 17, 116, 140 n.23, 144 n.14
Solomon, 43
Suppiluliuma I, 17, 188 n.504

Ancient Authors



Ancient Near Eastern Texts Cited

*Hittite*

Akkadian

*Biblical Texts Cited*

## Foreign Terms

*Sumerograms*



*Akkadian*



*Hittite*



*Hurrian*